The Life Beyond

The Life Beyond

The true story of a surgeon, his patients, and the supraconsciousness that proves the afterlife exists

Dr Manuel Sans Segarra

Juan Carlos Cebrián Barrientos
Translated by Dustin Langan

GODSFIELD

First published in Great Britain in 2026 by Godsfield Press, an imprint of
Octopus Publishing Group Ltd
Carmelite House
50 Victoria Embankment
London EC4Y 0DZ
www.octopusbooks.co.uk

An Hachette UK Company
www.hachette.co.uk

The authorized representative in the EEA is Hachette Ireland,
8 Castlecourt Centre, Dublin 15, D15 XTP3, Ireland (email: info@hbgi.ie)

Published in Spain by Editorial Planeta, Avenida Diagonal, Barcelona

ISBN 978-1-84181-684-5
eISBN 978-1-84181-686-9

A CIP catalogue record for this book is available from the British Library.

Typeset in 12/16pt Garamond Premier Pro by Six Red Marbles UK, Thetford, Norfolk.

Printed and bound in Great Britain.

1 3 5 7 9 10 8 6 4 2

Commissioning Editor: Louisa Johnson
Project Editor: Rimsha Falak
Copy Editor: Chris Stone
Art Director: Juliette Newsworthy
Production Manager: Sarah Parry

This FSC® label means that materials used for the
product have been responsibly sourced.

MIX
Paper | Supporting
responsible forestry
FSC® C104740

CONTENTS

I wish to dedicate this book to my parents, Manuel and Carmen, for their profound influence on my character and my professional journey, and to my wife, Magaly, and my sons, Jaime and David, who are so important in my life and whom I love deeply.
They have encouraged, motivated and guided me in writing this book.
I consider myself an incredibly fortunate person.

Dr Manuel Sans Segarra

To my grandparents, Paco and Isabel, who taught me the meaning of unconditional love, and who continue to watch over me from eternity. Before writing this book, I had a feeling it was true. Now, I am fully convinced.
To my parents, Pepe and Paqui, who gave me life and showed me the path to being a good person.
To my wife, Elena, and my daughters, Elena and Carla, who fill my days with joy. With them, I am walking the path of life. What has been the best part of your day?
To my sister, Noemí, her husband, Raúl, and my niece and goddaughter, Gabriela, with whom we share life's moments.
To my brother-in-law, Jordi Juez, and my friend, Juan Antonio Fernández. You triggered a series of serendipitous events that I will explain later, and of which this book is a part.
And, of course, thanks to life and for everything it gives me, day after day.

Juan Carlos Cebrián Barrientos

PREFACE

Dr Manuel Sans Segarra

Throughout my career as a surgeon, I have treated patients who were diagnosed as clinically dead from various causes. Their condition involved cardiac and respiratory arrest, areflexia (meaning no reflexive response to any stimulus) and a total absence of mental activity, as shown by a flat electroencephalogram (EEG) after 15 seconds. Some of these patients were successfully resuscitated and later described a series of events that had occurred during this period, known as a near-death experience (NDE). When following up with these patients, I observed the profound psychological impact their NDEs had on them, particularly on their outlook on existence.

My first encounter with these phenomena, the subject of a field of research called near-death studies (NDS), stirred a deep interest in me, as I wanted to understand their aetiopathogenesis (their causes and mechanisms) and their pathophysiology (the changes they cause in the body). I read the extensive literature on the subject and explored various related disciplines, such as neurology, psychiatry, psychology, theoretical physics, philosophy, metaphysics and religion. This deep enquiry into the human condition has undoubtedly shaped my own existential outlook.

It has shown me the reality of human life in its full depth and complexity. I believe our true purpose is to discover and live in accordance with *Supraconsciousness*[1], our existential reality, which makes us unique and unrepeatable and allows us to be happy and free. These insights have taught me how to manage my emotions better and to structure my life in alignment with archetypes.

As a specialist in general and digestive surgery, I approach my patients holistically, attending to their body, mind and spirit. I strive to cure or heal them and to alleviate their suffering when that is not possible. I also always seek to provide comfort to both the patient and their family.

I have verified that it is possible to connect with Supraconsciousness and thereby control the ego, our false identity, which I like to call the 'non-self'. We can achieve this by disabling the four powerful means at its disposal: ignorance, attachment to material objects, selfishness and fear. Ultimately, all fear is a fear of death.

I understand the vital importance of awakening the consciousness of others. The minimum proportion of the population needed to begin a shift in collective awareness is actually quite small. Therefore, we must achieve a critical mass of individuals who are aware of their existential reality. This is imperative to change the path along which we are headed, dominated by a prevailing egocentrism that ravages our atmosphere, hydrosphere, geosphere and biosphere, endangering both our civilization and our planet.

---

1 I have made an intentional distinction in the spelling here. *Superconsciousness* (from the Latin super, 'above') implies a higher level within one's own psyche, like an apex of intuition to which we can ascend. Whereas *Supraconsciousness* (from the Latin supra, 'beyond') implies a transcendent and non-local field that exists entirely beyond the individual mind.

PREFACE

Juan Carlos Cebrián Barrientos

I had not truly looked within until April 2023. Up until then, my life had been based on doing and having. Of course, my being existed, but it was uncharted territory.

I am 46 years old as I write this in January 2024, a journalist by trade and an entrepreneur by vocation. My life has brimmed with blessings and joy. I would even say that it has been easy, especially compared to the lives of my parents and grandparents.

Therefore, exploring my being has not simplified my life.

It already was simple.

However, thanks to this book and how it came about, which I will explain later, I now understand that life is not about earning more money, spending 15 hours in an office pouring over a business project or finding success and recognition, in the Western sense of whatever that may mean.

This does not mean that I reject money, that I am going to abandon my projects or that I suddenly no longer care about recognition.

I'm just putting those pursuits in their proper place. Both money and recognition must be understood as a consequence, but not as an end unto themselves.

There is a realization that is helping me tremendously. I want to share it with you.

Contrary to what I used to believe, what happens to us depends much less on our actions than we might like to think. What happens to us springs from our being, and not so much from our doings.

And our doings should relate to how we react to what life sets in our path, over and over. The value of our actions lies in how we face what life gives us and what it takes away.

In reality, what happens to us is caused by something deeper. It comes from what we feel, who we really are, how we are connected to the universe and how we relate to Supraconsciousness.

When you have finished reading this book, you will understand what that means. I have no doubt about it.

Fortunately, my grandparents and parents raised me to always trust in myself, and I did so. However, I never even considered what it meant to trust in life.

In this preface I want to tell you one thing: yes, you must always trust in yourself, of course, and yet the greatest achievement of all is to trust in life. If you take that quantum leap, extraordinary things will start happening to you. And by 'extraordinary things', I'm not referring to outward displays of success. That's not the point.

Success is merely a consequence. It is not the purpose of your being, nor should it be the purpose of your doings. If both are in harmony, you will live in peace and experience the feeling of happiness that we are constantly seeking.

It is a fine thing if success comes your way. And if it eludes you, that is perfectly fine too. In fact, we ought to reflect on what success truly means to us. If this book reaches millions of readers, can we consider that a success? And if just a single person in the entire world reads it and awakens their consciousness, can we consider that a success too?

Extraordinary things happened to me in 2023. Most of them were good, though some were painful. With time, I have come to understand that it is all part of life.

In closing, I want to explain something that has brought me peace and serenity: having had the enormous PRIVILEGE of knowing Dr Manuel Sans Segarra. And I write that word in capital letters, even if it may break some stylistic rule, because that is how I feel about it.

The first time that I heard him talk about Supraconsciousness and life after life was a turning point for me. Though I already knew something about NDEs, the importance of spirituality, physics and even quantum mechanics, my being needed to find Dr Sans Segarra for my consciousness to awaken. And for that, I give thanks to life.

In fact, when you read the chapter where I explain how I met him, you will understand that my actions had little to do with it. Instead, a chain of causal events, which I will describe in detail, had everything to do with it. And that is why I am fully convinced that there are no coincidences. There are only causalities.

This concept has looped back to me time and again these last few months, from people in totally unrelated circles.

PROLOGUE

Dr Mario Alonso Puig

I first met Doctor Manuel Sans Segarra at a conference to which we had both been invited to speak. What he shared and how he shared it had a profound impact on me. Throughout my career as a general and digestive surgeon in the United States and Spain, I had been fortunate to meet people who had survived NDEs and had read some books about the subject. However, though often treated as taboo or the product of an overactive imagination, I had never before heard anyone address it with such unyielding academic and scientific rigour.

Dr Sans Segarra and Juan Carlos Cebrián Barrientos imbue this incisive book with wisdom and philosophy, which they connect with the findings of quantum physics. Furthermore, regardless of one's personal beliefs, they provide compelling examples of people who have had NDEs and how they have transformed their lives.

Today, we must move past our dualistic and materialistic worldview. All too often we regard anything that cannot be labelled, weighed or measured with suspicion, and yet we witness with growing awe, wonder and gratitude how research is beginning to take more interest in the infinitesimal and the subtle. This is vital, for only those who can see the invisible can achieve the impossible.

Above all, this book is an act of courage, for only a brave soul could tackle such elusive subjects. It is therefore commendable that the authors have written such a lucid and engaging book about NDEs.

Albert Einstein, one of the fathers of quantum physics, argued that the separation between the inner and outer worlds is purely an illusion of the mind. I believe the time has come for us to explore, with humble curiosity, the extent to which matter and energy are not mutually opposed, but rather complementary. This magnificent book can help us on that journey.

1
A CHANGE OF PERSPECTIVE

They did not know it was impossible, so they did it.
Mark Twain

I am a physician specialized in general and digestive surgery, with a particular focus on oncological surgery. In this first chapter, I want to share my personal journey and how I came to change my perspective on life and consciousness.

My path to becoming a surgeon

I was the first physician in my family. My interest in medicine was sparked by my mother's stories about the Spanish Civil War, as she had been an operating theatre nurse, and by my father, who also worked in healthcare. I showed an innate interest in biology and anatomy from an early age and even once dissected a dead bird, curious to see what was inside.

My fascination with biology grew during secondary school, and by the time I was ready for university, it was crystal clear: I wanted to study Medicine. My parents supported my decision, and thanks to my good grades, I received a scholarship. However, what started out as a simple choice of degree gradually bloomed into a passion that led me to explore the depths of human life and beyond.

During my medical training, I realized that my true calling was surgery. I discovered two important aspects of this discipline: the practical part, which involves dissection and manual procedures, and the scientific part, which provides the theoretical basis for action.

In my third or fourth year of medical school, I began working in the emergency surgery department at Hospital Clínic in Barcelona, where I gained hands-on experience on call in the operating room. After graduating in Medicine and Surgery with excellent grades, I specialized in general digestive surgery under Dr Pedro Piulachs at the University of Barcelona's Faculty of Medicine for four years. Later, I expanded my theoretical and practical knowledge abroad before taking a permanent position at Bellvitge University Hospital in Barcelona.

Surgery allowed me to combine my love of biology with a deep desire to help others. Each operation was a unique challenge, requiring precision, skill and an encompassing understanding of human anatomy. But beyond surgical technique, I also learned about the importance of compassionate care and effective communication with patients. A physician must cure or heal, and alleviate suffering when that is not possible, but they must always provide comfort to the patient and their family.

My invaluable experience in France, the United Kingdom, Italy, Japan and the United States expanded my horizons and deepened my understanding of medicine. I worked with some of the world's top surgeons specializing in oesophageal and pancreatic surgery and had the opportunity to learn about the latest techniques and advances in digestive surgery.

As a surgeon at Bellvitge University Hospital, I had the privilege of using my skills and knowledge to improve my patients' lives. Each day I am reminded of why I chose this path and feel motivated to continue learning and growing as a physician.

The scientific approach

The Cartesian and Newtonian scientific method has been essential to my approach to medicine. It is based on the principle that every natural phenomenon can be explained by physical and mathematical laws. As such, I have dedicated my career to the pursuit of these laws within medicine, using the scientific method to test hypotheses and advance our knowledge.

As a surgeon, I have taken this same scientific approach in my clinical practice. I have used the latest research to inform my surgical decisions and to provide my patients with the best possible care.

As a university professor, I have had the opportunity to supervise residents' doctoral theses. This work has allowed me to guide the next generation of physicians and researchers, instilling in them the importance of critical thinking and a rigorous scientific methodology.

In short, my career has been guided by a commitment to the scientific method and a belief in the power of science to improve medicine. While this approach can be challenging, I am wholly convinced it is essential for advancing our understanding of human health.

A transformative experience

One day, during a shift in the emergency surgery department, my life took an unexpected turn. I resuscitated a patient from clinical death caused by a serious traffic accident. After I operated on him and his condition stabilized, the patient shared with me what he had experienced during that critical period.

What he told me led me to investigate and learn about NDEs in greater detail. I studied the work of experts in the field, such as Elisabeth Kübler-Ross, Raymond Moody, Eben Alexander, Melvin L Morse and others, and met with professionals from various disciplines to better understand these phenomena, including neurologists, psychiatrists and psychologists.

That experience changed my perspective on life and death. It made me question my previous beliefs and prompted me to explore areas of medicine and consciousness that I had never previously considered. I began to see that there was more to our existence than the Cartesian and Newtonian scientific methods can explain.

My patient's NDE showed me that there are parts of our consciousness that transcend our physical existence. This led me to study concepts such as *Supraconsciousness* and non-local consciousness.

As I went deeper into my research, I realized that these experiences were not as rare as I had previously thought. Many people have reported similar experiences. There are thousands of published cases and a growing body of scientific research supporting the validity of these accounts.

This journey has been challenging, but it has also been incredibly rewarding. It has expanded my understanding of what it means to be human and has enriched my medical practice in ways I could never have imagined.

Beyond the scientific method

Despite my scientific background, I concluded that the traditional scientific method could not fully explain NDEs. I sought answers in theoretical physics, which afforded me clearer insight into some of the phenomena reported by people who had these experiences.

Theoretical physics, especially quantum physics, provides a model of the universe beyond what we can perceive through our senses alone. This discipline suggests that reality is far more complex and mysterious than it may normally seem. Through my study of quantum mechanics, I began to appreciate that phenomena such as superposition and entanglement (which I will describe later) presented certain parallels with accounts of NDEs.

The reality of consciousness

Although I have never personally had an NDE, my research firmly convinced me that consciousness transcends matter, and that this could be objectively demonstrated through scientific methods. It was through practices like meditation and the exploration of non-local consciousness, however, that I experienced it directly.

Meditation, in particular, gave me access to states of consciousness beyond my everyday experience. Through these practices, I could directly experience the transcendental nature of consciousness in communion with the universe as a whole.

These experiences taught me that our consciousness is not limited to our physical body or our immediate sensory experience. Instead, it can tap into a broader field: Supraconsciousness or non-local consciousness.

In short, my journey from a purely scientific approach to a deeper, more holistic understanding of consciousness has been fascinating and utterly transformative. It has changed how I see the world and enriched my medical practice in unexpected ways.

A new view of existence

My outlook on existence was turned upside down. Before, I saw death as the absolute end, a logical conclusion according to materialist assumptions and the second law of thermodynamics. However, I now understand that our existence is eternal and transcends our body and mind.

I feel fortunate to have had this life-changing experience that radically upended how I see the world. Now I yearn to share this knowledge and help people to awaken to a deeper understanding of reality. This is why I share my perspective through lectures and recordings.

The importance of personal enquiry

I want to stress that what I say here should not be swallowed wholesale. Blind belief requires dogmas and leaders, while scepticism and doing your own research lead to the discovery of truth. I encourage those who read this book to be critical and to think, reason and study for themselves. Only by reaching your own conclusions can you truly enrich your understanding of life and consciousness.

As the philosopher José Ortega y Gasset said, 'A good teacher is not one who provides the audience with a wealth of conceptual knowledge, but one who, along with that wealth of knowledge, awakens critical thinking.' My goal is to make you think and question so you can find your own answers in your quest for truth.

2
INTRODUCTION TO SUPRACONSCIOUSNESS

A human being is a part of the whole, called by us 'Universe',
a part limited in time and space. He experiences himself,
his thoughts and feelings as something separated from the
rest – a kind of optical delusion of his consciousness.
Albert Einstein

Supraconsciousness, or non-local consciousness, is a fascinating topic that has captured the attention of many researchers and curious minds. This book explores the idea that consciousness is not merely a product of neuronal activity in the brain, but resides at a deeper, more fundamental level of reality. Most importantly, we will seek scientific foundations soundly supporting the existence of Supraconsciousness, for which references to quantum mechanics will be essential.

The primary aim of this book is to awaken consciousness and bring more happiness to its readers' lives. We understand that most of our readers will not be scientists specialized in quantum theory, which deals with phenomena related to subatomic particles, the smallest particles known. Therefore, we use the most accessible language possible and generally follow each technical explanation with a metaphor to aid understanding.

The idea that consciousness is more than just neuronal activity in the brain is not new. Since antiquity, philosophers and thinkers have debated the nature of consciousness and its relationship with the world around us. In recent decades, however, scientific research has begun to shed light on this topic, leading to a new understanding of consciousness that transcends conventional interpretation.

The relationship between consciousness and quantum physics is an ongoing area of research, driving intense discussion. Some scientists, such as the mathematical physicist Roger Penrose and the anaesthesiologist and psychologist Stuart Hameroff, for example, have postulated that consciousness is generated by quantum processes. According to them, the brain's neural system forms an intricate network and the consciousness it produces should obey the rules of quantum mechanics, the branch of physics that determines how tiny particles like electrons move. This theory ultimately suggests that consciousness could be the result of quantum processes occurring within brain cells.

They also propose that microtubules, which form the cytoskeleton of neurons, are structured in a fractal pattern that could enable quantum processes. Neurons are eukaryotic cells, meaning their nucleus, which contains the DNA, is separated from the cytoplasm by a membrane. Fractals are structures that are neither two-dimensional nor three-dimensional, but have an intermediate fractional value somewhere in between.

Penrose and Hameroff's conjecture is still considered highly controversial. Nevertheless, various researchers are conducting experiments to test some of the principles underpinning the quantum theory of consciousness. While there is still no definitive answer, these studies could bring us a step closer to understanding the complex relationship between consciousness and quantum mechanics.

Like a vast ocean

Imagine consciousness as an immense ocean. The waves on its surface represent the neural activity in our brains. Each wave is unique and fleeting, just like every thought and sensation we experience. However, while the waves are what we see and experience, much more lies underneath.

The vast depths of that ocean teem with currents and movements that we cannot see, but which are fundamental to the formation of the waves on the surface. These currents represent quantum processes, such as those proposed by Penrose and Hameroff, which could be happening inside our brain cells.

Furthermore, the ocean also exhibits fractal patterns, just like those found in the microtubules within our neurons. For example, ocean currents can flow in patterns that are neither completely two-dimensional nor three-dimensional, but something in between.

Just as the mysterious world beneath the ocean's surface continues to yield new discoveries, we still do not have all the answers about the relationship between consciousness and quantum mechanics. And yet, each new wave of research brings us one step closer to understanding the infinite ocean of consciousness.

We know of cases of patients who are conscious despite having minimal or no neuronal activity. This challenges the conventional understanding of the relationship between neuronal activity and consciousness and suggests that something more is at play.

There is a local consciousness that originates from the biochemical activity of neurons. We know this because when the activity of these cells is inhibited, what we call 'neuronal consciousness' is suspended. This is what happens during sleep, for example, or when a general anaesthetic is administered intravenously or by inhalation;

the substance reversibly stops neuronal activity and produces a loss of consciousness, thereby allowing for aggressive diagnostic or therapeutic intervention.

Consciousness provides us with awareness of our existence, our thoughts and our actions. At every moment, it allows us to know who we are, what we think, what we are doing and in what environment we are moving. As a consequence, it is accompanied by self-awareness or self-reflection, which can be compared to looking in a mirror to determine whether our decisions and actions are ethical or not. Conscience originates in the brain, but also in the social environment, as collective consciousness.

A fundamental property

Consciousness is a coherent state in which all parts act in unison. In higher states of consciousness, a transcendent bridge is formed between the material and the spiritual. This bridge generates a powerful sense of expansion and liberation, leading to peace, harmony and deep unity with nature and with the primordial energy of the universe. In these states, the ego is brought to heel and selfishness dissolves.

Supraconsciousness is the idea that consciousness is not merely the result of neuronal activity in the brain, but rather exists at a deeper and more fundamental level of reality. According to this view, consciousness is a fundamental property of the universe, present in all living and non-living things.

As the source of individual consciousness, Supraconsciousness unites us with the world around us. Although it may seem difficult to define, since it lies beyond our conventional understanding of reality, we can think of Supraconsciousness as a kind of energy field that permeates the entire universe. This energy is the source of individual awareness and is what allows us to experience everything within us and, most importantly, everything around us.

Throughout this book, we explore the idea that Supraconsciousness exists not only in our life, but also after our death and before our birth. This suggests that consciousness is eternal, transcending individual life and ever-present throughout the universe. The message is clear and deeply reassuring: death is not the end of consciousness, but simply a change in the way it manifests.

Beyond consciousness and death

The intuition that consciousness survives death has been explored by many cultures and religions throughout history. In some, individual consciousness is believed to reincarnate in a new form of life after death; in others, it is thought to merge with a universal or divine consciousness.

Supraconsciousness also invites us to reflect on the very nature of reality. According to this view, what is real and effective is not limited to what we perceive through our senses – there is much more beyond their grasp. A deeper, more fundamental reality underlies everything we experience, and this reality is the true source of consciousness. Supraconsciousness, therefore, also leads us to ponder the nature of both body and mind.

The mind and the body are not separate aspects of our being, but are interconnected and form part of a far greater whole. Supraconsciousness suggests that the mind and body are constituent parts of a broader system that encompasses the entire universe, and that consciousness is the force that unites them. In essence, each and every one of us is nature. We are stardust; we are universal energy.

In the chapters that follow, we will look more closely at NDEs, quantum biology and the relationship between the body's systems and the mind, among other subjects. But before we go further, it is important to have a clear understanding of what Supraconsciousness is and why it is worth exploring.

> Supraconsciousness goes beyond ordinary or normal consciousness: it is a state in which you perceive your communion with the entire universe and experience a sense of unity and wholeness. It cannot be grasped solely through thought or reflection, nor can it be fully understood using logic or reason. Instead, it is experienced directly through a profound sense of connection and interrelation with all existence.

Supraconsciousness is not reserved for a select few or those who have devoted their lives to meditation and spiritual practice, even though many may claim otherwise. It cannot be attained or achieved in a conventional sense; it is not a commodity that can be acquired or possessed. It is revealed when we open ourselves to it and allow ourselves to experience it directly. Supraconsciousness is available to every one of us. We need only be willing to explore and recognize it.

Words fall short when describing Supraconsciousness, for it transcends language itself. Therefore, any attempt at a definition is ultimately ineffective and incomplete. And yet, those who have experienced Supraconsciousness often speak of it in terms of a profound sense of peace, harmony, stillness, love and joy.

Views of consciousness throughout history

Different cultures and civilizations have developed their own theories and beliefs about the nature of consciousness and its relationship with the physical world. In fact, the study of both consciousness and Supraconsciousness is as old as humanity itself. From the earliest philosophers to modern scientists, it has been a fascinating subject of discussion across many fields of knowledge.

In the ancient civilizations of Mesopotamia and Egypt, which arose around 6,000 and 5,000 years ago, respectively, consciousness was believed to reside in the heart. The Egyptians considered the heart to be the centre of life and consciousness, the seat of a person's emotions, thoughts and character. This organ was so important to them that it was left intact in the body during mummification, while other organs were removed.

Ancient India (c. 7000–600 BCE) was a hotbed for the development of sophisticated theories about the nature of consciousness. Philosophers and thinkers in those times explored it deeply, leaving a valuable legacy that continues to influence our understanding of the concept today.

The Upanishads describe a state of pure consciousness known as Brahman, the universal soul. According to these sacred Hindu texts, written from around 800 BCE, Brahman is the ultimate reality and the source of all existence. This concept refers to a transcendental reality that is both immanent and transcendent.

In other words, Brahman is present throughout the universe, yet also exists beyond it.

Ancient Hindu traditions and philosophies also explored fundamental concepts such as Atman, or the eternal self. Atman was considered an individual's immutable essence, persisting regardless of any changes to the physical body and mind. Thus, it was held to be our true and constant self.

A central idea in the Upanishads is that Atman and Brahman are one and the same. This identification of the individual self with ultimate reality is a hallmark of Hindu thought. It is believed that realizing this unity leads to liberation (moksha) from the cycle of birth and death (samsara).

Ancient Greece was the birthplace of some of history's most influential philosophers, whose reflections on the nature of consciousness have left an indelible mark on Western thought. For instance, Plato and Aristotle (5th–4th centuries BCE) held contrasting but equally profound views on the subject.

Plato, a student of Socrates and teacher of Aristotle, believed that a realm of perfect and unchanging Forms existed beyond the physical world. According to his theory, our consciousness can access this realm of Forms through reasoning. Also known as the world of ideas, it is the home of eternal and immutable truths, which constitute ultimate reality.

Aristotle, on the other hand, viewed consciousness as intrinsically linked to the physical world. Unlike his teacher Plato, he did not believe in a separate realm of perfect Forms. Instead, he argued that our consciousness and understanding emerge from our interaction with the physical world. According to Aristotle, all knowledge begins with sensory experience.

During the Middle Ages, the interpretation of consciousness was heavily influenced by Christian theology. Spanning from the 5th to the 15th centuries, this was an era of profound reflection and debate about matters of faith, morality and the nature of humankind.

Saint Augustine of Hippo (354–430 CE), one of the most influential philosophers and theologians of that time, played a crucial role in shaping the medieval understanding of consciousness. Augustine viewed the human mind as an image of the Divine and believed that our capacity to think and understand was a way of participating in divine reason. In other words, he thought that our consciousness reflects divine knowledge and wisdom. Augustine believed that by using our ability to reason and comprehend, we can draw closer to God and understand His will. This view placed consciousness at the heart of the path to salvation and eternal life.

Augustine also closely linked consciousness with the concept of Original Sin, which, as told in the Book of Genesis, stemmed from Adam and Eve's disobedience when they ate the forbidden fruit in the Garden of Eden. This act divested all humankind of the holiness and righteousness that God had bestowed upon them. According to Augustine's interpretation, it is through our consciousness (and our conscience) that we become aware of our sinful nature and our need for redemption. This idea had a lasting influence on Christian teachings and remains a central theme in many Christian denominations today.

With the advent of modern science in the 17th and 18th centuries, the study of consciousness began to take a more empirical approach. This period saw the rise of great thinkers who changed our understanding of consciousness, with Descartes at the forefront.

One of the founders of modern philosophy, the French philosopher and mathematician René Descartes (1596–1650) is famous for saying, 'I think, therefore I am' (*Cogito, ergo sum,* in Latin). This short sentence, one of the best-known in the history of philosophical thought, is a cornerstone of the Cartesian system and places consciousness at the very core of our existence. Descartes argued that the only thing he could be certain of was his own existence as a thinking being. Though everything else might be questionable or deceptive, the fact that he was thinking was an undeniable certainty.

Descartes is also known for his theory of Cartesian dualism, which holds that the mind and body are two distinct substances. According to this theory, the mind (or soul) is a non-physical, thinking substance, while the body is a physical, extended substance. This view has had a lasting influence on our understanding of consciousness and remains a central point of contention in philosophy and science. Considered a father of the scientific method, along with the physicist, mathematician, theologian, inventor and alchemist Isaac Newton (1643–1727), who introduced calculus, Descartes believed that matter, the basic structural element of the universe, was governed by the laws of nature.

The concept of objectivity, introduced long before by Aristotle, presupposes a separation between observer and object, which is clearly a form of dualism. These principles could not be applied to mental activity, which was considered an epiphenomenon of the brain – something that accompanied it, but exerted no influence over it. Consequently, Descartes concluded that the mind should be studied by other disciplines, such as metaphysics, philosophy and religion.

The 19th century marked a major milestone in the study of consciousness, with the establishment of psychology as a scientific discipline. During this period, the study of consciousness became even more focused on empiricism, an approach based on observation and direct experience.

Often considered the founder of American psychology, William James (1842–1910) was key to solidifying this new, empirical approach. James described consciousness as a constant flow of thoughts and sensations, an idea known as the 'stream of consciousness'. He believed that our consciousness is not static or fragmented, but rather resembles a continuous and ever-changing torrent of thoughts, images, sensations and emotions.

James is also known for his functionalist approach to psychology. Instead of focusing simply on the content of consciousness (like thoughts or sensations), he was interested in understanding the purpose it fulfilled. For example, he wondered why we experience certain thoughts or emotions. He also asked how they help us to adapt to our environment.

The 20th century: a paradigm shift

The 20th century witnessed significant advances in our understanding of consciousness, driven by the emergence of new disciplines like neuroscience and psychoanalysis. Although many of these fields had deeper historical roots, the application of the scientific method and the march of technological progress enabled major breakthroughs across various fields that vastly expanded our knowledge of the human mind and Supraconsciousness.

Neuroscience, which seeks to understand the mechanisms our brain uses to produce phenomena such as consciousness, gained unstoppable momentum throughout the 20th century. Although we are still far from fully understanding exactly how our consciousness works, advances in neuroscience have provided some important clues. For instance, we now know that certain areas of the brain are associated with different aspects of consciousness, such as perception, attention and memory.

In the early 20th century, the neurologist Sigmund Freud (1856–1939) revolutionized our understanding of the mind by introducing the concept of the unconscious. Freud argued that our conscious actions and experiences are heavily influenced by unconscious desires and fears. This idea challenged the prevailing belief that we are entirely aware of our own motivations and behaviours.

Freud proposed a model of the human mind structured in three parts: 1) the id, which houses our primitive impulses; 2) the ego, which deals with reality; and 3) the superego, which acts as an internal moral critic. His theory posits that these three components constantly interact, thereby influencing our decisions and behaviours.

The theoretical physicist Albert Einstein (1879–1955), best known for his theory of relativity, also grappled with the concept of consciousness; his essays on science and religion provide valuable insight into this and related issues. In later decades, other researchers have explored possible connections between quantum mechanics and biological processes.

Einstein viewed the human being as part of the universe, and yet that it experiences itself as separate from the rest. This suggests a unified vision of consciousness and the cosmos, in which consciousness is not merely a product of the human brain, but an integral aspect of the universe itself.

Stephen Hawking (1942–2018), famous for his work in theoretical physics and cosmology, also took an interest in the study of consciousness. Hawking did not conduct formal research on consciousness, but he did publicly reflect on it from a scientific and computational perspective. For example, he was one of the scientists present at the signing of the Cambridge Declaration on Consciousness in 2012. Drawing on neuroanatomical, neurochemical and neurophysiological evidence, this document asserts that non-human animals have consciousness. This perspective broadens the concept of consciousness beyond humans to include other living beings.

In recent decades, neuroscience has provided new ways to explore consciousness. Advances in technology have enabled scientists to observe the brain in action and begin to unravel how its countless neurons work together to produce our conscious experience.

Roger Penrose and Stuart Hameroff have posited a theory linking consciousness to quantum physics. Their model, called Orchestrated Objective Reduction, or Orch OR, suggests that microtubules, structures formed by the protein tubulin inside brain cells, could support quantum processes, potentially contributing to the formation of consciousness.

Despite these advances and novel theories, the study of consciousness remains a formidable challenge. Consciousness is inherently subjective, meaning that each person has their own unique and non-transferable experience. This makes it difficult to study objectively and quantitatively.

Furthermore, though we know that certain areas of the brain are associated with specific aspects of our conscious experience, we still do not fully understand how subjective awareness arises from these cerebral processes. This mystery is known as the 'hard problem of consciousness'.

Nevertheless, scientists are making steady progress in our understanding of consciousness. And with each new discovery, we move closer to answering some of the most profound questions about who we are and our place in the universe.

Prepare to embark on a journey that will forever transform your understanding of life.

3
NEURONAL ACTIVITY AND CONSCIOUSNESS

I regard consciousness as fundamental. I regard matter as derivative from consciousness. We cannot get behind consciousness. Everything that we talk about, everything that we regard as existing, postulates consciousness.
Max Planck

From a physiological perspective, consciousness could be defined as a state of the nervous system that enables the emergence of complex, conscious behaviours that depend on temporary neuronal operations predominant in certain regions of the brain. Such complex behaviours, which may be thoughts or actions involving some type of movement, could previously only be measured through direct observation of behaviour. But today, thanks to technological progress, we can take a 'snapshot' of the brain activity related to each task using devices that capture functional neuroimages.

The scientific approach applying the laws of physics

Consciousness possesses two key properties that help us to understand how it works: our level of alertness and our experience of awareness.

Our level of alertness corresponds to our degree of physical and mental activation at any given moment, meaning the amount of energy we expend to stay attentive to our organism's immediate needs. When we are asleep, drowsy or under anaesthesia, our level of alertness drops so low that our consciousness effectively 'switches off'. In this state, our brain and body need adequate physiological activation to process the information arriving through our senses.

To understand this better, we can compare our consciousness to a lighthouse. When the lighthouse is lit (high alert level), it illuminates the surroundings, allowing us to perceive and respond to stimuli. However, when the lighthouse is off (low alert level), our ability to perceive and respond to stimuli diminishes or vanishes entirely.

Research suggests that neuronal activity arises primarily from a network of neurons located in the medial frontal regions and the posterior cingulate cortex of the brain. This network is the neural basis of conscious activity. Imagine it as an orchestra: each neuron is a musician playing its instrument – the microtubules within the neurons – in the symphony of our consciousness.

In summary, while the laws of physics and traditional scientific methods dictate that there is still much to learn about consciousness and its relation to neuronal activity, current research suggests that it results from complex interactions between different brain regions. This raises a crucial question: is the brain merely permissive of consciousness, acting as a passive conduit, or is it productive, meaning it actively generates it?

Life from a traditional scientific perspective

In the vast universe of science, the Cartesian and Newtonian scientific method stands as a beacon of light, guiding researchers through the shadows of the unknown. This approach, which has been the cornerstone of scientific enquiry for centuries, applies to both physical and emotional problems and encompasses everything from the deepest mysteries of the cosmos to the most intimate challenges of human life.

Through this scientific lens, life is perceived as a finite phenomenon. This view is a direct consequence of the fundamental principles of the scientific method. To illustrate this, we could think of life as a book. Every book has a beginning and an end. While the pages in between may be filled with unexpected twists and turns, the story always concludes on a final page.

In summary, the Cartesian and Newtonian scientific method provides us with a framework for examining and understanding the world around us. It allows us to address both physical and emotional problems rigorously and systematically. And while this method may lead us to see life as finite, it also reminds us that there is much we do not know about it, and that boundless possibilities for exploration lie ahead.

The Cartesian and Newtonian scientific method

Articulated by René Descartes and Isaac Newton in the 17th and 18th centuries, the scientific method has proven to be a powerful tool for discovering the laws of nature. This approach seeks to accurately reproduce natural phenomena in a controlled laboratory environment.

The fundamental principles of the scientific method include duality or objectivity – a concept introduced by Plato and later developed further by Aristotle – and material monism, which holds that matter is the fundamental substance of nature. This method is also based on the principle of locality, which posits that there is always a cause-and-effect relationship with the reality of objects, and that these objects can be defined by objective parameters.

This process can be visualized as the work of a watchmaker taking apart a watch to understand how it works. Each part of the watch, each gear and each spring, has a specific place and function. Like watchmakers, scientists deconstruct natural phenomena into their most basic components (particles of matter) to understand how they assemble the universe as we know it. This is material reductionism, a concept introduced by Democritus, the Greek philosopher who founded atomism around the 4th or 5th century BCE. According to atomism, the universe is composed of indivisible particles, called atoms, that move in a void.

The Cartesian and Newtonian scientific methods also maintain that all movement has continuity, with a beginning and an end.

These principles of the scientific method lead to a clear determinism, casting human beings as mere observers in the universe. Since everything depends on natural laws, we cannot influence them.

Albert Einstein said: 'Science without religion is lame, religion without science is blind'. This quote reflects the interdependence between science and other forms of knowledge. Although the scientific method can provide a deep understanding of how the universe functions, it also recognizes its own limitations. Science can describe how things work, but not necessarily why they exist in the first place.

Thus, the scientific method studies the macroscopic universe, grounded in materialism. This establishes a separation, or chasm, between science and metaphysics, between matter and spirit. Descartes held that the scientific method was contingent on the study of natural laws and matter, leaving transcendent phenomena to other disciplines such as metaphysics, philosophy and religion.

The scientific method is materialist and realist. It operates through observation and experimentation to seek out and understand natural laws and thereby reproduce them in the laboratory.

Theoretical physics complements the scientific method. The antagonism between matter and spirit is not real; it has its parallel in the particles and waves of the quantum field. The scientific method addresses the macroscopic universe, while quantum mechanics focuses on the microscopic.

In summary, the Cartesian and Newtonian scientific method provides a systematic and rigorous framework for exploring and understanding the natural world. Although this approach may seem limiting to some, it also opens up infinite possibilities for discovery and understanding, as discussed above.

Human beings from a scientific perspective

With its fundamental principles of objectivity and material monism, the scientific method provides us with a lens through which we can examine and understand human nature. According to this approach, human beings are composed of body and mind. The body is matter, which is tangible and observable. The mind, however, is more ethereal, composed of feelings, emotions, memories, consciousness and learning.

Despite lacking any material basis from the viewpoint of the scientific method, these mental phenomena are considered epiphenomena. This means that they are considered a consequence of the neuronal metabolic activity in our brain.

We can imagine this process as a river. The water (matter) can be seen and measured, but the currents flowing beneath the surface represent mental phenomena (thoughts, feelings and experiences), which are less visible and more difficult to gauge. However, these currents are still an integral part of the river and have a significant impact on its behaviour and direction. Eric R Kandel, a neuroscientist who won the 2000 Nobel Prize in Physiology or Medicine, has spent much of his career studying these phenomena. Kandel explores how our experiences shape our neurons and how these neurons then structure our minds. By asserting that the mind is what the brain does, he highlights the relationship between complex brain activity and mental experience, as well as the need to continue deepening the study of how to measure and quantify the processes, memories, feelings and learning that reside in our minds.

The relationship between physical death and existence

To summarize, René Descartes was one of the most influential philosophers in history. His famous phrase, 'I think, therefore I am', prompts us to reflect on the nature of existence. According to this perspective, our capacity to think and reason is what defines us as human beings.

If our thoughts, feelings and experiences are like the currents that shape our metaphorical river, then when these currents stop (that is, when we stop thinking), the river ceases to flow. In this sense, physical death, which involves the cessation of neuronal activity, marks the end of our existence. António Damásio explores this idea in his book *Descartes' Error*. This Portuguese neuroscientist argues that our feelings and emotions are essential to our sense of self, so Descartes was mistaken in separating the body from the mind. Although these phenomena can be difficult to measure and quantify, they are an integral part of what makes us human. Therefore, Damásio suggests that our ability to think is rooted in our state of being.

The 'marvels' of the mind are inseparable from the structure and function of the organism. This is because the brain and the rest of the body constitute a set of biochemical and neural regulatory circuits that interact with the environment as an indivisible whole. Therefore, Damásio concludes, mental activity arises from this relationship.

The theoretical physicist Richard Feynman, the co-winner of the Nobel Prize in Physics in 1965, reflects on this idea in his book *Surely You're Joking, Mr. Feynman!*, which is based on various taped conversations he had with a friend. In one such chat, he argued that even though death is inevitable, we should not fear it. Instead, we should celebrate life and all that it offers.

Death also raises questions about the meaning and purpose of life. Why are we here? What is our purpose? These are questions that have intrigued philosophers and scientists for centuries. The biologist, ethologist and zoologist Richard Dawkins addresses these questions in *The Blind Watchmaker*. Dawkins argues that even though life may appear to have a purpose, it is actually the result of natural processes with no direction or design, making it impossible for the evolution of life to depend on the will or action of a creator.

The process of dying occurs in two stages:

1) The neocortex disconnects, which causes disorientation and confusion. This engages our midbrain, or limbic system, where our entire personal history is stored. The dying person may experience moments of pleasure and suffering, which can cause agitation as they relive these memories.
2) The basal or reptilian brain, which autonomously controls vegetative functions, becomes compromised. This leads to disturbances in heart rhythm and respiratory apnoea.

Death is not the enemy of life, but a part of it. We start dying from the moment we are born. We should interpret life as a preparation for death. Death allows us to return to our origin.

There are five stages in the conscious response to death. These stages can occur separately, overlap or coincide:

1) Upon learning of a life-threatening illness, the patient first responds with surprise, denial and isolation or withdrawal.
2) This is followed by a stage of active denial, characterized by anger, rage and resentment. The patient refuses to accept what is happening to them and sees it as an injustice meted out by fate.
3) Faced with the evident reality, they try to strike a bargain. This does not mean true acceptance, but it is rather an effort to postpone the inevitable.

4) This is the most conflict-ridden and sorrowful stage for the ill or dying person, who sinks into a depression and feels a profound sense of loss. Sometimes, they are so overwhelmed with despair that, seeing no solution, they even take their own life. In my professional experience as a physician, I have sadly seen cases of patients diagnosed with cancer who, upon learning of their serious condition, decided to die by suicide. Healthcare professionals must be acutely aware of the dangers of this stage to properly understand and support the patient.
5) Finally, patients accept their situation and fight actively and positively, holding onto hope for a potential cure.

Despite these reflections, death remains a mystery to many of us. Although the scientific method can provide some answers, it also reminds us how much we do not know.

What does quantum theory tell us?

Quantum theory, which deals with phenomena at the atomic and subatomic scale (as we will see in Chapter 6), has provided a new perspective for understanding consciousness. Some scientists believe that consciousness is generated by quantum processes. Though controversial, this idea has opened new avenues of research and debate in the field of cognitive neuroscience.

In addition to Roger Penrose and Stuart Hameroff (see page 20), other researchers continue to explore the possible relationship between consciousness and quantum physics. In an effort to test this relationship, Xian-Min Jin, a professor at Shanghai Jiao Tong University, has investigated with his team how quantum particles might move within a complex structure like the brain, but in a laboratory setting. If their findings can one day be compared with measured brain activity, we could be a step closer to validating or disproving the Penrose–Hameroff theory.

In recent decades, there has been a growing interest in understanding and demonstrating the relationship between consciousness and underlying neural activity. As a result, many studies and experiments have been conducted using animals and patients with brain damage. By combining genetic and optical methods, and by harnessing the remarkable precision of laser light to 'target' specific groups of neurons, we now have neuroimaging systems that allow us to observe what happens in the deep brain structures and networks related to consciousness.

Consciousness and quantum mechanics

Quantum mechanics is a branch of physics that deals with the behaviour of particles on a very small scale, such as atoms and subatomic particles. The concept was introduced by the German physicist and mathematician Max Planck (1858–1947), who received the Nobel Prize in Physics in 1918 for demonstrating that energy depends on wave frequency and the universal constant (h) that bears his name. Simply put, the higher the frequency, the greater the energy. Planck also demonstrated that electromagnetic waves travel in packets, which he called 'quanta' (from the Latin *quantum,* meaning 'quantity'); it is from this term that the name of quantum mechanics is derived.

At this scale, particles can exhibit strange and unexpected behaviours, such as being in multiple places at the same time (superposition) or affecting each other instantly, regardless of the distance between them (entanglement).

To better understand this, we can think of quantum superposition as a musician playing several instruments at once. In the macroscopic world, this would be impossible. But in the quantum world, a single particle (the musician) can exist in multiple states (playing several instruments) simultaneously.

Quantum entanglement, on the other hand, could be compared to two perfectly synchronized dancers. No matter how far apart they may be, if one changes their movement, the other will respond instantly and change theirs too.

Subatomic particles are not visible, but we can detect their effects and interactions. The language of quantum mechanics is mathematics, meaning that we can express it via mathematical formulas. This provides us with a new paradigm for interpreting the subatomic world.

The impact of the quantum explanation

The idea that quantum processes could play a role in consciousness is known as 'quantum consciousness'. According to this explanation, famously defended by Penrose and Hameroff, quantum phenomena such as superposition and entanglement could occur within the brain and contribute to the formation of consciousness. In line with this theory, the microtubules within the brain's neurons could act as quantum channels, thereby allowing quantum processes to take place.

If validated and explored, this quantum explanation would completely transform our understanding of consciousness, life, death and the universe itself. The fascinating impact of the quantum explanation can broadly be summarized in four key consequences:

1) We could better understand what consciousness is and how it functions. This may lead to advances in both neuroscience and psychology.
2) If we could understand how consciousness is linked to quantum processes in the brain, we may be able to develop new medical treatments for disorders of consciousness, such as coma or Alzheimer's disease.
3) If our consciousness is a quantum process that can exist independently of our physical body, this may fundamentally change our view of life and death and the relationship and transition between them. The quantum explanation could provide a scientific basis for NDEs and out-of-body experiences.
4) If our consciousness can exist independently of our physical body, this raises profound philosophical and ethical implications and questions about our identity, our purpose and our place in the universe.

4
ENIGMATIC PHENOMENA

Science is the belief in the ignorance of experts.
Richard Feynman, *The Pleasure of Finding Things Out*

Near-death experiences (NDEs) occur in patients diagnosed with clinical death, which can arise from various causes. Clinical death is defined as a state in which the patient experiences cardiac arrest, indicated by a flat electrocardiogram (ECG) with an isoelectric line confirming the absence of a heartbeat; respiratory arrest; areflexia, or a total lack of tendon reflexes; and no mental activity, with a flat EEG after 15 or 20 seconds.

In my experience as a surgeon, the most frequent causes of clinical death that I witness are trauma, suicide attempts, the ingestion of toxins, the administration of drugs or anaesthetics and haemorrhages or serious complications during surgical procedures. However, there is no doubt that the highest incidence of clinical deaths in medical services is due to cardiac and cerebral vascular accidents.

We can compare clinical death to a computer that has shut down. Just as a computer without power cannot perform any functions, a human body in a state of clinical death shows no signs of life.

However, just as a computer can be restarted, some patients can be brought back to life through cardiopulmonary resuscitation

(CPR). This technique is similar to pressing the power button on a computer to restart it: if done quickly, it can bring the system back to life. As a result of current scientific progress, up to 33 per cent of patients can be revived if CPR is applied within the first minute after cardiac arrest. However, this proportion drops to approximately 14 per cent after the first minute.

These figures underscore the importance of acting swiftly in medical emergencies. Each second counts when it comes to saving a life.

Sam Parnia, a physician, resuscitation researcher and associate professor of medicine at the NYU Langone Medical Center, is one of many researchers working to unravel these mysteries. In his book on what he calls *The Lazarus Effect* – named after the biblical account of how Jesus Christ resurrected his friend Lazarus, who had been dead for four days, by saying, 'Lazarus, come out!' – Parnia explores NDEs and what they can teach us about life and death.

Parnia begins with an introduction to NDEs, explaining that they have been reported by people close to death. These individuals often describe the phenomena as involving encounters with a bright light or sensations of immense peace and love. Through his research, Parnia seeks to understand what happens in the brain during such experiences and how they might change not only our understanding of life and death, but also how we have traditionally defined them.

Parnia's book also explores the scientific advances that have allowed physicians to revive people who have been clinically dead for up to 7 hours. For instance, he presents the case of a woman who suffered a cardiac arrest while in hospital. Doctors were able to revive her after 45 minutes of CPR. Although the woman had been clinically dead during that time, she later described an NDE in which she found herself in a place filled with light and peace, with specific patterns that I will describe in the following chapters.

This case, and many others like it, raises questions about what it truly means to be alive or dead. How is it possible for someone to have a conscious experience while clinically dead? What does this tell us about the nature of consciousness and its relationship with the physical body?

Parnia uses cases such as the above throughout the book to explore these questions from a scientific perspective. He argues that NDEs and advances in resuscitation are challenging our traditional concepts of life and death and suggests that we need new ways to understand them.

His pioneering work on the Lazarus effect is a fascinating attempt to unlock the secrets of NDEs and the insights they provide into the ultimate nature of our existence. Readers are therefore invited to explore these phenomena from a scientific perspective and consider their profound implications for our understanding of life and death.

NDEs can vary as much as those who have them. Some report feelings of peace and unconditional love, while others explain that they have been reunited with deceased loved ones. There are even those who describe leaving their bodies and watching others struggle to save their life from an outside perspective. Regardless of how they recall these NDEs, a strikingly similar pattern always emerges.

These experiences remain an enigma to modern science. Although we know that they occur, we do not fully understand why or what they mean. After studying them in depth, I have observed a certain parallel between quantum phenomena and the occurrences reported by patients during NDEs.

Despite being shrouded in mystery, NDEs have a profound impact on those who have them. In fact, many report lasting changes in their attitudes and beliefs afterwards. Some even say that they have partially or fully overcome their fear of death and that they feel a greater appreciation for life.

Characteristics of NDEs

NDEs are actually surprisingly common. The phenomenon is widely documented, with thousands of cases described in the scientific literature around the world. These experiences follow a distinct pattern with common features, where the differences are more a matter of nuance than purely conceptual.

Patients who have had an NDE are firmly convinced that they are completely distinct from dreams. They remember it for the rest of their lives in vivid detail, and it usually has a profound psychological impact on their existential outlook, changing the path they tread and the roles they assume for the remainder of their 'new' life.

References to NDEs can be found throughout human history. Plato describes how a Pamphylian soldier named Er was grievously wounded in battle. Though initially presumed dead, he recovered and described an NDE.

The incidence of NDEs does not differ based on sex, age, race, socioeconomic status or religious belief. However, individuals' cultural background and religious beliefs do influence their account of their NDE, as they often find language inadequate to describe the variety and intensity of these occurrences. Not all patients who recover from clinical death have NDEs. The incidence in various studies ranges between 18 per cent and 25 per cent. However, the actual incidence may be higher, as patients who have had these experiences do not always share them openly due to fear of being misunderstood, so they go unreported.

NDEs with frightening content have been described. They are not common, with an incidence below 5 per cent. Such patients feel dragged into darkness, experiencing terrifying visions and feelings of guilt. These cases involve intense and prolonged emotional trauma that may require psychological support. They are also known as hellish NDEs. Their cause has not been determined, though patients who have them have sometimes had tumultuous lives.

Bruce Greyson, a professor of psychiatry at the University of Virginia, defines NDEs as profound subjective experiences with a transcendent or mystical component that occur in people near death. They involve the sensation of leaving the body, transcending the ego, space and time.

In the early 1970s, Raymond Moody gained worldwide fame for his research on people who were 'resuscitated'. A psychiatrist and professor emeritus of psychology, with a PhD in Philosophy, Moody defines NDEs as conscious experiences that occur during life-threatening situations.

Patients describe the following main recurring patterns of NDEs:

1) They often describe it as hyperreal or 'more real' than everyday reality.

2) Such patients may have an out-of-body experience, meaning they leave their physical body and can observe it and what is happening around them from an outside perspective.
3) Some report extraordinary perceptions, such as hearing what others are thinking. There are clinical cases in which the patient has been able to recount events happening at that very moment in other locations, or events that had yet to take place.
4) Those who have had an NDE sometimes later describe moving through a dark tunnel or void towards a bright light.
5) They may feel they have entered a different dimension, often described as a particularly beautiful and peaceful place.
6) Some patients report meeting other beings, which may include deceased loved ones or others who are sometimes referred to as 'guides'.

Case 1

A patient diagnosed with clinical death due to severe trauma describes her own NDE to Dr Sans Segarra

What I first noticed was that I was leaving my body and rising up in the air. I felt bewildered, not understanding what was happening to me. My surprise only grew when I saw my body on a stretcher, with medical staff trying to resuscitate me.

I wanted to communicate with the medical staff, but I couldn't speak. Then I tried to touch you, but shockingly, my hand passed right through your body. Later, I also passed through the wall and could observe everything that was happening in the different parts of the Emergency Department.

Gradually, I began to feel joy and peace.

I met beings that I would describe as 'beings of light', who helped me to stay calm. My greatest delight was when I came into contact with my dead mother. I also heard soft and very pleasant music.

After passing through a dark area, I saw a bright light that had a great pull on me. As I drew near, I was overcome by a calm and blissful feeling of pure love, unlike anything I had ever experienced before. Words cannot possibly express it.

I saw images of my life, as if they were playing on a screen.

In my ecstasy, the beings of light told me that my human life cycle was not over and that I had to return to my body. I balked, since I didn't want to lose such rapturous feelings, but they forced me.

My re-entry into my body was quite rough.

Case 2

An NDE with unique features in a patient undergoing treatment for oesophageal cancer

At that time, all surgery was open surgery. The patient was scheduled for an oesophagectomy with an intrathoracic gastroplasty (a Lewis procedure) to treat a squamous cell carcinoma of the mid-thoracic oesophagus. During the procedure, a major haemorrhage occurred at the aortic arch, which sent the patient into cardiac arrest. The haemorrhage was controlled and direct cardiopulmonary resuscitation was applied to the heart until his pulse was finally restored.

After surgery, the patient told me: 'At one point, I became aware that I was leaving my body. From above, I could see you massaging my heart and using the defibrillator. The operating room staff were rushing around. I experienced a feeling of peace and harmony. Finally, I returned to my body.'

Patients' descriptions of NDEs share these characteristics:

- Communication with other beings is only possible through thought.
- They only experience the present moment. There is no past or future.
- They can move unfettered by space and time simply by thinking about it.

Many other transcendent phenomena challenge our current understanding. These include telepathy, clairvoyance, precognition (the ability to see or perceive future events), mystical experiences, psychokinesis (the ability to influence objects, and even move them, through mental concentration), reincarnation and deathbed experiences. Though less common than NDEs, these phenomena also raise intriguing questions and are documented by clinical cases.

Hypnosis as a therapeutic tool

I should preface this section by clarifying that, unlike the topics discussed previously, I am not writing here from personal experience. I have no direct background in hypnosis. Nevertheless, I have decided to introduce readers to the subject and present the most prominent researchers who have explored it.

As a therapeutic technique, hypnosis can be used to achieve temporal regression, helping people to explore traumas and emotional or psychological problems that may stem from past experiences. During regression, the therapist guides the patient into a state of deep relaxation and focused concentration. This allows them to access their subconscious and explore memories that may originate from previous lives.

A leading figure in this field is Brian Weiss, an American physician and psychiatrist whose research and theories focus on reincarnation, past-life regression and the survival of the soul after death. Since the 1980s, Weiss has conducted regression therapy with thousands of patients at his practice in Miami, Florida.

Based on his work, Weiss argues that the existence of past lives can be substantiated by cases that demonstrate certain characteristics.

These include speaking a foreign language that one has never learned or been exposed to, known as xenoglossy; meeting people in this life who are the children one had in a previous life, who subsequently confirm details of those shared experiences; and recalling specific information during regression that was previously unknown and can later be independently verified, including places, dates or names.

> Regression therapy makes use of tools such as hypnosis, relaxation and visualization to help patients to trace the origin of their unresolved issues to the unconscious mind. The root cause is often identified in past lives, whose distant events reflect the symptoms appearing in the present.

Weiss has also written about messages relayed through his patients from highly evolved, non-physical entities that he calls 'master spirits'. He interprets these messages as providing valuable insights that can help his patients to address their current problems.

One of the most famous cases in Brian Weiss's work is that of Catherine, a young woman who came to his office seeking help to alleviate her recurring nightmares and anxiety attacks. During therapy sessions, Weiss used hypnosis to bring her into deep relaxation. Once in this state, Catherine began recalling details of what appeared to be past lives. These memories were so vivid and richly described that even Weiss was taken aback. Most astonishingly, Catherine even revealed facts about Weiss's life that she could not possibly have known.

Though Weiss's theories have sparked controversy in the scientific community, his research has had a significant impact on the field of psychiatry and has opened new avenues for exploring human consciousness and understanding life and death. Past-life regression remains a controversial technique, but many patients have reported significant benefits after undergoing it.

5
UNANSWERED QUESTIONS

The eternal mystery of the world is its comprehensibility . . .
The fact that it is comprehensible is a miracle.
Albert Einstein

As a disciple of the conventional scientific method, I arrived at a dead end, unable to find sound answers or explanations to the crucial aspects of NDEs. And not only was I unable to make sense of NDEs, but I was left equally perplexed by the entire sweeping landscape of transcendental phenomena.

I regarded this all as an unfathomable mystery with no material basis in fact. Each of these phenomena was a loose thread in the fabric of our understanding of the universe, threatening to unravel the very norms and conventions upon which I had built my career.

My training as a surgeon had always made me comfortable with anatomy and physiology, but here the scientific method provided me with no conclusions. When faced with any organic or emotional phenomenon, physicians try to understand two fundamental aspects: the cause (aetiopathogenesis) and the intimate mechanism of production (pathophysiology).

I sought help from and met with experts in other disciplines who were more knowledgeable about physiology and brain activity, such as neurologists, psychiatrists and psychologists. I told them about my patients' NDEs, indicating that I had no scientific explanation for them.

They were all quick to tell me the same thing: 'Those are hallucinations.' Hallucinations, they said, are a series of mental manifestations produced as a consequence of a profound metabolic disorder that neurons undergo due to a lack of blood flow in the brain.

The importance of oxygenated blood

Though the brain only accounts for 2 per cent of our body weight, it consumes 20 per cent of all our energy. The source of energy powering the brain is glucose. However, the brain does not accumulate reserves of it, unlike the muscles, where it is stored in the form of glycogen.

Neurons are highly sensitive to a lack of oxygen (anoxia) or glucose in the blood, but also to an increase in carbon dioxide (hypercapnia), potassium (hyperkalaemia) or acid levels in the body (metabolic acidosis). All these states, which occur in circumstances such as cardiac arrest, can end up causing clinical death in patients. Injuries caused by the deprivation of blood, and therefore oxygen, are irreversible after 5 to 10 minutes. That is why cardiorespiratory resuscitation (CPR) must be practised as soon as possible.

The parts of the brain most sensitive to a lack of oxygen are the cerebral cortex, hippocampus, thalamus and basal ganglia. The brainstem, which controls vital involuntary functions such as the heart rate and breathing, can better withstand a lack of oxygen.

A lack of blood flow destroys neuronal synapses (the spaces that link the end of one neuron with the next or with another cell) and the cell membrane (the true brain of the cells). It also triggers a disruption in the levels and flows of minerals, the formation of free radicals (naturally occurring substances that can cause irreversible damage when they proliferate) and the destruction of proteins.

Possible causes of NDEs according to the scientific method

Several hypotheses based on the scientific method have been proposed to explain NDEs, framing them as hallucinations:

- NDEs are produced by the brainstem to avoid the pain of terminal death, as a type of feigned-death strategy used by less complex animals. This hypothesis is easy to disprove due to the 'sophisticated' experiences that can reportedly occur in NDEs when the neocortex is non-functional.
- They are psychedelic visions caused by an administered drug. This is also easily refutable, as drugs interact with neocortical receptors (which are not functioning) and many patients have no history of drug use.
- Another hypothesis proposes that near-death experiences (NDEs) may result from an intrusion of REM sleep phenomena into wakefulness or into a critical physiological state. During REM sleep, the brainstem plays a central role in generating vivid dream activity while simultaneously inhibiting motor neurons, producing the characteristic muscular paralysis associated with this sleep phase. REM sleep is also associated with complex neurochemical

mechanisms involving neurotransmitters such as serotonin and acetylcholine, which act as chemical messengers regulating neuronal activation and inhibition. These neurotransmitters interact mainly with cortical neural networks involved in ordinary conscious experience. However, this hypothesis does not seem sufficient to explain the full phenomenology of NDEs. During clinical death, the neocortex is considered functionally inactive, and many reported NDEs include highly structured conscious experiences, intense clarity, lasting memory retention and profound psychological transformation. This is why REM intrusion may account for certain partial features of NDEs, but it does not adequately explain their overall complexity and consistency.

- They are attributed to DMT (N,N-dimethyltryptamine), a chemical compound secreted by the pineal gland, due to stress on the brain. Like serotonin (5-hydroxytryptamine), DMT can trigger extremely intense hallucinations. This hypothesis is also unsound because hallucinogens affect the neocortex, which is not operational during NDEs.
- They already exist in the deep regions of the limbic system, in the lateral amygdala, and are activated when the neocortex disconnects, much like a computer rebooting.
- They are caused by the action of the neurotransmitter glutamate, whose effect is similar to that of ketamine, a hallucinatory anaesthetic. However, such hallucinations are unpleasant and chaotic, in stark contrast to typical NDEs.
- They are explained by the preservation of some small part of the neocortex that does remain active, despite the widespread

poor perfusion (little or no supply of blood and oxygen) in the brain.

- They are ascribed to psychotic and especially schizophrenic outbreaks or to drug addiction. However, the patients in question have no prior history of either.
- They are described by some psychologists as a sophisticated defence mechanism to cope with the catastrophe of clinical death.
- They are attributed to endorphins and enkephalins, opioid neurotransmitters released to counter the stress experienced when close to death, causing feelings of tranquillity and happiness.
- They are provoked by cerebral anoxia, which, by potentially causing neuronal disinhibition, might tip the neurological balance, making the neurons 'go crazy' and generating frenetic activity similar to that which produces seizures.
- They are linked to a reduction in gamma-aminobutyric acid (GABA), a neurotransmitter that has an inhibitory effect on neurons, potentially provoking great neuronal excitability.

Why NDEs are different

When we compare hallucinations with the manifestations of NDEs in patients diagnosed with clinical death, clear differences emerge:

1) NDEs follow a coherent narrative structure, whereas hallucinations are meaningless and absurd. Furthermore,

NDEs often have recurring features that differ more in terms of nuance than conceptually. In contrast, hallucinations vary completely from one patient to another, so they have nothing in common.

2) Patients remember every last detail of their NDE, even years later. People who experience hallucinations forget them quickly and may even feel ashamed to talk about them.
3) NDEs have a very profound psychological impact on patients, and especially on their existential outlook, leading to a change in their approach to life. No such 'transformative' effect occurs after a hallucination.

The scientific community clearly agrees that there is a great clinical difference between hallucinations and NDEs.

The scientific method also fails to explain how a series of biochemical reactions between molecules – which are made of particles, and therefore matter – within neurons can give rise to a mental response, a thought or consciousness, none of which has any material basis.

NDEs involve scientifically inexplicable phenomena, such as the ability to pass through solid structures with ease. Another astonishing phenomenon is the ability to provide meticulous descriptions of situations occurring simultaneously in another location, and even on the other side of the world. Such a transfer of information, independent of space and time, only makes sense if it is somehow directly witnessed, at least with the current state of our knowledge.

Functional magnetic resonance imaging (fMRI) scans have been performed on patients' brains while they discussed their NDE in painstaking detail. The use of this advanced tool, which measures tiny changes in blood flow in the brain, as well as changes in metabolism and neuronal activity, has allowed us to observe the following results:

- The prefrontal cortex is activated. This is the region of the brain where the highest-order cognitive functions are located, such as rational thought, decision-making and free volition. This provides clear evidence that NDEs have a coherent narrative structure.
- When patients describe situations that carry a significant emotional weight, activation is reflected in the temporal lobes.
- The occipital lobe activates when they discuss images that particularly captured their interest. The recall of visually striking objects is facilitated by the involvement of mirror neurons. The activation observed when a patient describes these objects during an fMRI scan indicates memory retrieval in the mirror neurons. This suggests that the patient genuinely saw the object in question; in other words, they are not lying.

The scientific method cannot provide the aetiopathogenesis and pathophysiology of NDEs. All attempts to do so are merely unproven hypotheses. What is clear is that there is some phenomenon behind our local or neuronal consciousness that we do not understand and that lies beyond the reach of scientific enquiry.

We have covered a body of evidence that fully supports the authenticity of NDEs, ruling out the possibility that they could be hallucinations, dreams or figments of the patients' imagination.

6
OPENING OUR MINDS TO QUANTUM PHYSICS

Anyone who is not shocked by quantum theory has not understood it.
Niels Bohr, Werner Heisenberg's memoir
Physics and Beyond

Physical death, which occurs with that final exhale that wrests us from the Earth's embrace, draws the curtain over our body and therefore over our cerebral consciousness. However, like a persistent echo rippling behind the silence, there is another kind of life, holding fast after physical death. This life can shed light on objective phenomena, demonstrated through empirical investigation. Until recently, we were stuck in a labyrinth with no way out and unable to understand why.

Then, like someone struggling to light a lamp in the darkness, I delved into the murky yet fascinating world of quantum physics. I had read studies suggesting that this branch of science could help to unravel the mystery, so I sought out its custodians, the quantum physicists, and explained my conundrum. I told them we cannot discover the truth about NDEs with the scientific method and asked if we might approach it from a quantum perspective.

The quantum physicists regarded me with surprise and admitted that they knew little about NDEs. However, they were more than willing to help me to understand the basic principles of quantum physics. As Isaac Newton said, 'I seem to have been only like a boy playing on the sea-shore . . . whilst the great ocean of truth lay all undiscovered before me'. And so, I set out on this new adventure, hoping that these principles might illuminate this phenomenon that the scientific method could not explain.

The quantum physicists warned me that I would have to change my approach and abandon materialist ontology to understand their field. Richard Feynman implied that quantum mechanics is beyond anyone's understanding. This always reminds me of a famous, and perhaps apocryphal, anecdote I like to share:

At a meeting between Albert Einstein and Charlie Chaplin, the scientist is said to have praised the actor for his ability to communicate so effectively without words, as seen in his silent films, where audiences could understand even without dialogue. Chaplin reportedly turned the compliment around by admiring Einstein's ability to talk at length about complex subjects that few people truly understood. Heeding their warning, I prepared for a complete paradigm shift. To navigate the deep waters of quantum physics, I would have to leave the solid ground of the scientific method and materialist ontology behind.

A brief overview of elementary particles

Particle physicists use accelerators to propel subatomic particles nearly to the speed of light. This makes them collide with each other, allowing scientists to study their fundamental structure and behaviour. These collisions also generate new particles, enabling analysis of the composition of matter at its most basic levels.

Fundamental particles include electrons, quarks and bosons (such as gluons and photons). There is no hierarchy between quarks and photons. They fall into different categories and neither is more fundamental than the other. Quarks form the atomic nucleus, while photons are bosons that carry light.

These are the fundamental particles and their properties:

- Electrons are elementary particles with a negative electric charge.
- Quarks form protons and neutrons, the building blocks of atomic nuclei.
- Bosons mediate three of the fundamental forces of the universe.
- Photons carry the electromagnetic force, allowing light to propagate.
- Gluons bind quarks through the strong nuclear force.
- W and Z bosons govern the radioactive decay of some particles.
- The Higgs boson drives the mechanism by which fundamental particles gain mass by interacting with the Higgs field.

  The Higgs boson was proposed by Peter Higgs and other physicists in 1964 and its existence was confirmed experimentally at CERN in 2012. This discovery gave us a better understanding of the mass of elementary particles and was recognized with the Nobel Prize in Physics in 2013. The term 'God particle' has been popularized by the media, but it is not accepted by the scientific community, as it has connotations unrelated to physics.

- The fourth fundamental force, gravity, has not yet been incorporated into quantum physics. A hypothetical particle called the graviton has been proposed as its carrier. However, it has not been detected or integrated into a complete theory thus far.

It must be stressed that these particles operate at the most foundational level of the universe and are responsible for the properties and behaviours of all matter.

Therefore, we can conclude that:

- The basic constituent of the universe is not matter, but energy. Everything is energy.
- Energy is neither created nor destroyed; it only changes form.
- Energy travels in waves of varying frequency and intensity.

Basic quantum principles

The French theoretical physicist Louis-Victor Pierre Raymond de Broglie, the 7th Duke of Broglie (1892–1987), made a remarkable discovery while studying subatomic particles: they could be in several places at once and could manifest either as a wave (energy) or as a particle (matter). He concluded that both were two extremes of the same element: energy.

In his PhD thesis, published in 1924, de Broglie postulated the wave nature of electrons and argued that all matter has wave properties. This idea, known as the 'de Broglie hypothesis', laid the foundation for what we call wave-particle duality and became a central pillar of quantum mechanics. This wave-particle duality is explained by the principle of superposition.

These ideas were so revolutionary that some physicists rejected them at first, but Albert Einstein supported them enthusiastically. De Broglie was the first high-profile scientist to call for the creation of a multinational laboratory, leading to the inception of CERN.

His hypothesis was finally confirmed by the electron diffraction experiment, conducted by the American physicists Clinton J Davisson and Lester H Germer in 1927, which demonstrated that electrons can indeed behave like waves.

For his contribution to quantum physics, Louis de Broglie was awarded the Nobel Prize in Physics in 1929. However, his work on quantum physics is not limited to his hypothesis. It was developed into a full theory of wave mechanics by Erwin Schrödinger (1887–1961), a physicist and mathematician who helped to create quantum mechanics and formulated the probability wave equation that bears his name.

The Danish physicist Niels Bohr (1885–1962), who won the Nobel Prize in Physics in 1922, took de Broglie's ideas even further. Based on Thomas Young's (1773–1829) famous double-slit experiment, Bohr made a fascinating observation. When a stream of electrons passed through the double slit, they appeared on the screen as waves with interference patterns. However, when a conscious observer monitored the electrons passing through the slits, they transformed into particles. Building on the work of de Broglie, Niels Bohr emphasized that the act of measurement and the experimental setup determine whether quantum phenomena appear with wave-like or particle-like behaviour. This phenomenon is called 'wave function collapse', which can be understood as energy collapse.

Later thinkers such as John von Neumann, Fritz London, Edmond Bauer and Eugene Wigner explored interpretations in which conscious observation could play a role in the collapse of the wave function.

In 1939, the physicists Fritz London and Edmond Bauer proposed that conscious observation plays a role in completing the quantum measurement process. Drawing on phenomenological ideas, they suggested that consciousness and the observed system participate jointly in the emergence of definite physical reality. John von Neumann and Eugene Wigner further explored the possible relationship between consciousness and quantum measurement, although these interpretations remain debated within modern physics.

This suggests a revolutionary idea: human beings can collapse energy. If so, we are no longer merely observers of the universe, but rather co-creators of our reality.

Living beings do not merely observe the world around us; we also help to create our own reality. The determinism of the scientific method is not absolute. Everything material we observe is an illusion; it is energy collapsed into matter. Eastern philosophies such as Hinduism and Buddhism reached the same conclusion centuries before Christ through meditation and spirituality. Both belief systems posit that everything we perceive is an illusion and encapsulate this concept in the Sanskrit term *maya,* though with different meanings. In

Hinduism, *maya* is the creative force that conceals Brahman, or true reality. In Buddhism, it is a deceptive aspect of the mind that casts an illusion over our perceivable surroundings.

During a famous meeting in 1927, a debate arose as to whether quantum mechanics was based on causality or probability. Einstein defended causality with his famous remark that '[God] does not play dice'. However, Bohr, Schrödinger and Werner Heisenberg backed the dynamics of probability . . . and won.

Heisenberg (1901–1976) won the Nobel Prize in Physics in 1932. Together with Schrödinger and Paul Dirac, he described the uncertainty principle. Heisenberg postulated that certain atomic properties, such as the position and velocity of the electron, could never be known with absolute certainty because they cannot be measured simultaneously.

Quantum entanglement was first highlighted in a 1935 thought experiment by Albert Einstein, Boris Podolsky and Nathan Rosen. Decades later, John Stewart Bell (1928–1990) derived a theorem that allowed experimental tests of entanglement. In a typical entanglement experiment, a pair of related particles (such as photons or electrons) is separated, often across large distances. When a quantum property of one particle is measured (such as its spin), the outcome is instantaneously correlated with the measurement outcome of the distant particle. This correlation persists regardless of distance. This seemed to contradict Einstein's theory of relativity, which holds that no signal can travel faster than the speed of light (approximately 300,000 km/s). Together with Podolsky and Rosen, he formulated the EPR paradox to argue that quantum mechanics must be incomplete. They proposed that there must be hidden variables (i.e., unknown factors) that would explain the correlations without requiring instantaneous action at a distance.

Quantum entanglement does not involve the transmission of information between particles. Rather, it reveals correlations between measurement outcomes that cannot be explained by classical local physics. This phenomenon was formalized through Bell's theorem and later experimentally confirmed by Bell-test experiments such as those conducted by Alain Aspect.

Certain parameters of the scientific method change in quantum mechanics. Time is linear in the scientific method, with a past, present and future. In quantum mechanics and relativity, time is a complex concept that depends on the frame of reference. Einstein demonstrated that time is relative and Hawking explored models in which it could behave differently from our classical perception. Only the present truly exists – the current moment, right now.

The Schrödinger equation allows us to model the behaviour of quantum systems and, under appropriate conditions, classical mechanics emerges as an effective approximation. In this sense, Newtonian mechanics can be understood as a limiting case within the broader framework of quantum physics, valid at macroscopic scales and under specific conditions.

Some researchers in consciousness studies and NDEs have proposed that consciousness may not be entirely dependent on ordinary local brain activity. In this context, reports of awareness during periods of severely diminished or absent measurable brain function have led to hypotheses regarding a possible non-local aspect of consciousness, which in this book is referred to as Supraconsciousness.

The big takeaway

If we structurally apply the basic quantum principles to human beings, just as was done earlier with the principles of the scientific method, this analogy gives rise to four main conclusions:

1) Some theoretical perspectives – including bioelectromagnetic theories, quantum models of consciousness and research into neural oscillatory dynamics – suggest that the human body can also be understood as a complex system of electromagnetic interactions and oscillatory processes operating at different frequencies. Every material object is made of atoms and is actually mostly empty space. To grasp this scale, imagine one atom of your body with a nucleus the size of a golf ball placed at the centre of a football pitch. The electrons would be spinning in the furthest stands, and even further away. The spaces between subatomic particles are immense.
2) Quantum tunnelling, a phenomenon in which particles can probabilistically cross energy barriers according to quantum mechanics, has been proposed in some theoretical biological models, but it is unjustifiable within the framework of classical Cartesian mechanics. It is generally observed at small scales, where subatomic particles, which exhibit wave-particle duality, can overcome barriers under certain conditions that a classical particle could not, such as passing through a wall, for example. It is as if there were a tunnel allowing the particle to pass through a barrier of higher energy than its own.
3) The mind, with all its mental activity, has a clear quantum explanation. Emotions, feelings, thoughts, memories and

local consciousness are high-frequency electromagnetic energy, which we can register on an EEG. Due to superposition, this energy can manifest as matter or as a wave.

4) Given the evidence, we must accept that Supraconsciousness exists. It is a subtle, high-frequency energy that persists despite clinical death and has continuity outside the brain. This subtle energy, a non-local consciousness, accounts for what patients report after having an NDE.

7
AS IN LIFE, SO IN DEATH

The unexamined life is not worth living.
Socrates

Our physical body is a manifestation of low-frequency energy. After physical death, our body ceases to exist in the form we know, but this does not mean the end of our true existence.

Our existential reality, which is non-local consciousness, persists beyond our physical death. This non-local consciousness is our true essence and is not restricted by time and space as we experience in our physical reality. Indeed, this form of existence transcends the limitations of our physical body.

After our body dies, our non-local consciousness continues on its journey beyond the physical plane. Though our physical body has stopped functioning, the energy that composes it does not disappear. According to the principle of conservation of energy, it transforms and endures in other states.

Therefore, I say again: although death spells the end of our physical body as we know it, that is not the end of our existence.

The structure of matter

Our physical body is composed of matter. Democritus, a Greek philosopher who lived between 460 and 370 BCE, proposed that all matter was composed of tiny particles called atoms. Democritus's principle was widely accepted for centuries. The atom was believed to be indivisible, meaning that no smaller particle could exist.

However, advances in quantum mechanics have shown us that this view is incomplete. Though the atom remains a fundamental part of the structure of matter, we have discovered that it is composed of even smaller subatomic particles.

These subatomic particles include protons, neutrons and electrons. Furthermore, as we have seen, protons and neutrons are composed of even smaller particles called quarks. Thus, Democritus's view that the atom is indivisible has been replaced by a more complex and nuanced model of the structure of matter.

The entire universe tends towards entropy

Our physical body obeys the laws of thermodynamics after death. In particular, it is governed by the second law of thermodynamics, which states that everything in the universe tends towards entropy.

Entropy is a concept that refers to the natural tendency towards disorder or chaos. In the absence of laws or forces that maintain order, every system tends to move towards a state of greater disorder. This is what happens to our body after death: without the vital functions that maintain physical order, it decomposes and returns to the natural cycle of life. In other words, since our body is borrowed stardust, it returns to the universe via decomposition.

However, while our physical body is subject to these thermodynamic laws, our non-local consciousness is not. Supra-consciousness, as previously discussed, is a form of existence that transcends the physical limitations of our body. Although our physical body may decompose and succumb to chaos, our non-local consciousness abides.

How order is maintained in our life

Over the entire course of our physical existence, laws govern the order within our organism and the perfect biology of our body. And yet what is it that upholds these laws, enabling our body to develop in an orderly fashion throughout our life? The answer is our authentic essence: the primordial energy within each of us, our lifeforce, which is non-local consciousness.

Non-local consciousness acts upon local consciousness, influencing how we perceive and experience the world. Though we are grounded in our physical bodies, our non-local consciousness allows us to transcend these physical limitations and attune to a broader field of awareness.

This interaction between non-local and local consciousness can be seen as a dynamic dance. The non-local, with its broader perspective and engagement with the whole, guides the local, helping it to navigate the physical world. At the same time, our local experiences inform our non-local consciousness, letting us grow and develop as conscious beings.

Hence emerges the hypothesis that the brain is permissive, whereas non-local consciousness is productive.

Microtubules and information transfer

While investigating the effects of anaesthesia separately and unaware of each other's work, Penrose and Hameroff arrived at the same conclusion regarding a phenomenon called synchronicity. Both discovered that the relationship between non-local and local consciousness occurs in structures called microtubules.

In the 1990s, thirty years before Penrose won the Nobel Prize in Physics for demonstrating that black hole formation is a robust prediction of Albert Einstein's general theory of relativity, both researchers joined forces to propose an ambitious theory of consciousness. According to their theory, the brain's neuronal system forms an intricate network and the consciousness it produces should obey the rules of quantum mechanics.

This theory, known as Orchestrated Objective Reduction (Orch OR), suggests that consciousness arises from quantum vibrations in microtubules, which are protein structures that form part of the cytoskeleton of neuronal cells. These structures play a crucial role in various cellular functions, such as cell division and intracellular transport. In the context of consciousness, Orch OR proposes that microtubules could act as channels for transferring information between non-local and local consciousness.

Imagine a cell is a bustling city. Microtubules would be the underground network of this large metropolis. Just as underground trains shuttle people from one place to another, microtubules transport molecules and vesicles, the micro-components responsible for storing or digesting cellular products and waste, to different parts of the cell.

And just as a well-designed underground system helps to keep a city organized and running smoothly, microtubules help to preserve the cell's structure and ensure that it functions efficiently. Furthermore, much like the tunnels of the Tube or subway enable travel beneath the bustling surface, microtubules might allow our consciousness to travel beyond the physical limitations of our body.

The transformation of the body

After death, the vital systems that stave off disorder break down and the decomposition of the body begins.

However, this gradual decomposition entails no loss. Though our physical body transforms as it breaks down, not a single atom vanishes and not even the tiniest measurement of energy dissipates. Instead, it all transitions, becoming part of a different arrangement of energy and matter.

We can think of this as a metamorphosis, similar to the process by which a caterpillar becomes a butterfly. Despite drastic changes to the physical form, the underlying essence, in this case consisting of atoms and energy, persists and simply becomes something new.

Therefore, even though physical death is the end of our body as we know it, our existence is not extinguished. Our atoms and our energy endure in new forms, perpetuating the infinite cycle of life and death.

Fearless before death

When we die, our non-local consciousness, which is our authentic essence, is released and migrates to a different plane in a new energetic state.

This transition can be viewed as a journey to a new stage of existence, beyond the physical limitations of our body. As the body may return to dust and disorder, our non-local consciousness remains intact.

Therefore, we can conclude that we have no reason to fear death. Death is not the end, but a transformation and a shift towards a new form of existence. Rather than fearing it, we can view it as a natural and inevitable part of our existence, another step in our journey as conscious beings.

We can embrace death as an integral part of life, a necessary step in our eternal journey as conscious beings. It is not an end, but a transformation and a new beginning.

So why are we afraid?

It is understandable that many people fear death for various reasons. One is that the transition from life to death, and everything preceding it, can be traumatic, painful and unpleasant. Dying can involve great loneliness and anguish.

Understandably, one of the factors that most contributes to the fear of death is our instinct for self-preservation and our drive to stay alive. This tremendously powerful impulse is what makes us cleave to life. In reality, it is merely a biological means to safeguard the organism's physical survival amid dangers and risks.

To die is to take a step into the unknown, provoking fear, insecurity and anxiety. Finally, death means leaving behind everything we have achieved and fought for throughout our lives, including our family, our friends and our material possessions. Here I would like to say that instead of enjoying life, we often waste our limited time accumulating things that we cannot take with us when we die. We come into this world with nothing and we leave it the same way.

Nevertheless, it is important to remember that while death may be hard, it is not the end of our existence. Let me assure you again: our non-local consciousness, which is our true essence, continues beyond our physical death.

How would we live if we understood the true reality of existence?

If we fully understood existential reality, we would probably act differently. However, we live in a society that often promotes ignorance, poverty, illness and fear.

Such ailments may be perpetuated by various powerful forces, including civilization in general, political structures and the media. These forces often have an interest in miring people in ignorance and fear, as their subjugation makes them easier to manipulate and control.

However, it is critical to bear in mind that each of us has the power to seek the truth and to question the narratives presented to us. By grasping our true nature as conscious beings and acknowledging our ability to transcend the physical limitations of our bodies, we can begin to shake free from fear and empower ourselves to live more authentically.

We know that our physical body decomposes after death. It was never anything more than a garment or a casing. The atoms that compose it return to the universe and can become part of other matter. Though these atoms were lent to us for a time, they must eventually be returned.

As the saying goes, 'As in life, so in death.' Therefore, instead of fearing death, we can choose to live as fully and authentically as possible.

Living in harmony

Those who live while aware of their existential reality tend to die peacefully, harmoniously and joyfully. They are not afraid of death because they understand that the cycle of birth, love and death encompasses the three basic principles of our life.

By understanding and accepting these principles, we can enjoy our lives more fully and meaningfully. There is no need to fear death, as it is simply part of the natural cycle of life. Instead of fearing it, we can embrace it as a transition to a new stage of our existence.

So, instead of living in fear of death, we can choose to live in alignment with our existential reality. We can celebrate life, love deeply and face death with equanimity and acquiescence. This is the key to a rich and purposeful life.

8
THE PROPERTIES OF SUPRACONSCIOUSNESS

When an inner situation is not made conscious,
it happens outside, as fate.
Carl Jung, *Aion: Researches into the Phenomenology of the Self*

For those of us who believe in physics, the distinction between past, present and future is only a stubbornly persistent illusion.
Albert Einstein[1]

Supraconsciousness, as a non-local, subtle, high-frequency energy, exhibits the following properties:

- It is timeless, as some interpretations associated with modern physics and the philosophy of time suggest that the distinction between past, present and future may not be as absolute as it appears in everyday experience. In quantum terms, eternity is the absence of a past or a future.

1 This idea is inspired by reflections expressed by Albert Einstein in a 1955 letter to the family of Michele Besso, where he suggested that the distinction between past, present and future may be 'a stubbornly persistent illusion'.

- It is holistic with respect to universal quantum consciousness. The philosophical term 'holistic' embraces a given field or concept as a whole, such that the properties of the whole are not merely those of its constituent parts, but rather those parts themselves possess the properties of the whole. The parts are similar but may exist at different scales. This relates to the mathematical concept of a fractal, defined by algorithms and studied by the mathematician Benoît Mandelbrot (1924–2010), a professor at Yale University. A fractal is a geometric object in which the same structure is repeated at different scales and orientations, over and over again.
- It is a hologram. Much like a laser emits a pure, intense light to create a hologram of a three-dimensional object, an image with no real depth, so universal quantum consciousness is a hologram composed of all individuals' own Supraconsciousness, such that each contains the properties of the whole. The properties of universal quantum consciousness are omnipresence, omniscience (here we might think of intuition, which always expresses the truth), omnipotence and eternity. As the scientist and inventor Itzhak Bentov (1923–1979) said in his book, *Stalking the Wild Pendulum: On the Mechanics of Consciousness* (1977), 'So here we are, all part of this great hologram called Creation, which is everybody else's self, the Supraconsciousness of everybody else. It's all a cosmic play. And there is nothing but you!'
- It is part of the whole, lovingly united with the entire universe. We are the universe. We are nature. Here, the concept of dualism or separation between observer and object, which characterizes the scientific method, vanishes completely.

- It is responsive. The hologram of the universally merged Supraconsciousness enables the fulfilment of our most sincere wishes and prayers at the very moment we utter them.
- It is interconnected. Through the hologram of Supraconsciousness, any change in our life is reflected in each and every part of our world. This explains the synchronicities that occur despite physical separation.
- It is our authentic identity, which makes us unique and unrepeatable. It has the ability to collapse energy into matter and it is holistic with respect to universal quantum energy. In other words, Supraconsciousness is the presence of the primordial energy within each of us.

Many have elaborated on our communion with universal quantum energy. The French writer Victor Hugo (1802–1885) considered it to be the presence of God in each of us. In his book *Autobiography of a Yogi*, a highly evolved Hindu guru named Paramahansa Yogananda (1893–1952) described it as the finite manifestation of the Infinite within every being. And how many times did Jesus Christ say that human beings, children of God, are made in the image and likeness of Abba, the Father?

Aspects of Supraconsciousness

Supraconsciousness cannot manifest in three-dimensional form, as it is a high-frequency energy, but it does express itself in different ways that attest to its existence:

- **Introspection and deep meditation** allow us to perceive that our individuality does not emanate from our body or our mind. Beyond them lies Supraconsciousness, the source of our unique and authentic identity.
- **Intuition** is the expression of the omniscience of our Supraconsciousness. When we are faced with a problem, we rationally search for possible solutions through the activity of the prefrontal cortex in the frontal lobes of the brain, often without drawing a clear conclusion about what to do. However, our intuition bypasses reasoning and spontaneously provides us with the right solution. It never leads us astray. The more spiritually awakened someone becomes, the sharper their intuition.

  Intuition is a form of knowledge that transcends logic and rationality. It is an immediate perception or understanding that circumvents reason. It is often described as a 'gut feeling' or a 'sixth sense' that reveals information or the answers we seek. Intuition can arise as a hunch, a premonition or a visceral sensation.

  We can view intuition as a manifestation of the subtle, high-frequency energy of Supraconsciousness. According to the French philosopher and writer Henri Bergson (1859–1941), known for his work on the philosophy of mind, intuition is a tendency that develops from Supraconsciousness. Through intuition, we can perceive metaphysical oppositions, such as space and duration, matter and life, necessity and freedom and intelligence and intuition, as inversions of Supraconsciousness.

- **Creativity** is an expression of non-local consciousness at a given moment in time by means of painting, music, sculpture or any other kind of artistic language.
  Creativity is a manifestation of Supraconsciousness that enables us to create something new from 'nothing'.
  This capacity to create is not limited to the production of physical objects, but also includes ideas, solutions to problems, methods, interpretations and artistic forms. Creativity is the freest form of self-expression and can reflect and help to sustain an individual's emotional health.
  This is why works of art are unrepeatable. When the Post-Impressionist painter Paul Gauguin (1848–1903) was asked how he came up with his paintings, he is said to have replied that he shut his eyes in order to see. Machines will never create. What they really do is manage data at inhuman speeds, using algorithms and software that are products of human ingenuity.
- **Transcendent experiences**, including NDEs, originate in non-local consciousness, or Supraconsciousness. These include, but are not limited to, telepathy, clairvoyance, precognition, mystical occurrences, psychokinesis, reincarnation and deathbed experiences.
- **Archetypes**, as discussed by Plato and especially by the psychiatrist and psychologist Carl Gustav Jung (1875–1961), are universal principles that govern the thinking of all humankind and indicate whether our actions are ethical. They are an expression of our Supraconsciousness and shape a positive life dynamic dominated by altruism,

empathy, kindness, justice and especially love, which many philosophical and spiritual traditions have considered to be one of the most transformative force in human experience. Immanuel Kant, the 18th-century German philosopher and one of the most influential thinkers of his time, thought that there are two major realms of reality: an external one, which we observe in the grandeur and perfection of the universe, referring to the primordial quantum energy; and a second, internal one, which evaluates whether or not our actions are ethical and moral, experienced as an inner feeling of peace and harmony or of unease. They are the expression of our non-local consciousness, which is holistic with primordial quantum consciousness.

- **Happiness.** We will never attain happiness unless we act in accordance with our Supraconsciousness. Our materialist-based identity is externally defined; it is changeable, insecure and fearful. It is our ego, which I define as the 'non-self', and it depends largely on the opinions of others. Its goals are material in nature, including recognition, success, fame, wealth, power and dominance. To achieve them, the ego frequently succumbs to negative life dynamics characterized by excessive competitiveness, mistrust, jealousy, hatred, aggression and violence, which the politician and thinker Mahatma Gandhi (1869–1948) defined as the limit of incompetence. The decay in values, aggression and violence in Western society are a consequence of its prevailing egocentrism.

  The ego dwells on the past, causing feelings of guilt, and projects into the future, triggering anxiety due to uncertainty. It is uneasy in the present, our authentic existential reality. It moves between opposites, with a constant struggle at the boundary: life-death, health-illness, happiness-suffering, light-darkness and so on. The ego strives to eliminate the opposite that does not suit its interests, but this is never possible, since one produces the other, as is so aptly expressed in the *Tao Te Ching* (c. 4th century BCE), often attributed to the Chinese philosopher Lao-Tse. If we draw a curved line, it is concave on one side and convex on the other; one cannot exist without the other. Opposites are an illusion of the ego.

Pleasure, induced by the ego, is an intense emotional response to an external stimulus. In the long run, this recurrent response can form a habit, which requires increasingly greater stimuli. Often, pleasure is followed by suffering. This is due to the action of dopamine, a neurotransmitter that stimulates neuronal activity. The ego separates and isolates, creating a clear dualism between the observer and the object. Loneliness causes anguish and easily leads to depression. Antidepressants are among the most frequently prescribed medications today. True happiness originates from within us, in the Supraconsciousness, and brings peace, harmony, joy, stillness and silence. It is a response to a neurotransmitter, serotonin, which can inhibit and slow down neuronal activity.

- **True freedom of choice** is a property of Supraconsciousness. We will never be genuinely free under the control of the ego. Our life dynamics are shaped by our character, which we could define as our way of being, thinking and acting. Our character conditions our thoughts and our thoughts, in turn, condition our feelings.

Aristotle, whose words are just as relevant today, masterfully expressed how character is structured. I paraphrase him thus:

Thought conditions action. Action determines behaviour. Repeated behaviour creates habits. Habits structure character. And character marks destiny.

We define a habit as our mind's automatic response after a period of learning. This mechanism has the great advantage of saving us a vast amount of mental energy in our daily lives. Starting from the age of 35 or 40, for example, over 90 per cent of our actions are habits.

When do we acquire the habits that structure our character? The process begins in the womb and continues for the first seven to nine years. Parents are often scarcely aware of the tremendous importance of this stage of life. Habits are acquired primarily in the home and, later, in society at large.

Our decisions and reflections depend entirely on our character and therefore on our free will. If we are to be free, it must be our true identity, Supraconsciousness, that makes our decisions.

It is not easy to be authentic and truly free, as allegedly expressed by Thales of Miletus (624–548 BCE), one of the pre-Socratic philosophers:

The most difficult thing in life is to know oneself, and the easiest is to speak ill of others.

Can we change our habits and therefore properly structure our character if it is not authentic? The answer is certainly yes.

Neuroplasticity

The Spanish scientist Santiago Ramón y Cajal, who won the Nobel Prize in Physiology or Medicine in 1906, introduced the concept of brain plasticity, which he summarized with the phrase 'Every man, if he is so determined, can become the sculptor of his own brain'. This concept was not understood or accepted in his time. Decades later, following progress in neuroscience, the concepts of neuroplasticity and neurogenesis proved Ramón y Cajal right. In 2000, Eric Kandel received the Nobel Prize in Physiology or Medicine for his contributions to our understanding of neuroplasticity.

Neuroplasticity is the central nervous system's capacity to form new neural connections. This process is enhanced by neurogenesis, the generation of new neurons from neural stem cells, if the nervous system is appropriately stimulated. This effect is so significant that neurophysiologists contend that we can keep our brain adaptable and resilient for our entire life. Neuroplasticity and neurogenesis are induced by three fundamental pillars:

1) Healthy living, including a balanced diet, both in quality and quantity, as well as regular physical exercise.
2) Intellectual activity, which may come in various forms and degrees, such as reading, studying, playing a musical instrument, travelling, engaging in memory games and so on.
3) Social interaction, as engaging with others and our surroundings triggers an energy exchange that stimulates brain activity.

The Tibetan Buddhist experiment

In 2005, a group of scientists asked the current Dalai Lama, the spiritual head of Tibetan Buddhism, for permission to evaluate spiritually evolved Tibetan monks using EEG monitoring. They proposed that the monks would enter states of deep meditation capable of producing voluntary out-of-body experiences. The Dalai Lama agreed, as he is a proponent of quantum concepts. The monks displayed normal readings under baseline conditions: delta waves (0.1–4 Hz) during deep sleep, theta waves (4–8 Hz) in the twilight stage upon waking, alpha waves (8–12 Hz) during imaginative activities and beta waves (12–30 Hz) during the waking phase. All these types of waves, classified by their amplitude, are of low, medium and high frequency. Some studies on advanced meditation practitioners have reported unusually sustained and highly synchronized gamma-wave activity (30–100 Hz) during deep meditative states, which researchers associated with heightened states of awareness and neural integration.

Together with findings from advanced meditation research and consciousness studies, NDEs have contributed to renewed scientific and philosophical discussions about the nature of consciousness and the possibility that awareness may not be entirely reducible to ordinary brain activity.

9
QUANTUM BIOLOGY

Life seems to be ordered matter that evades rapid decay
into the thermodynamical equilibrium of death.
Erwin Schrödinger, *What is Life? The Physical Aspect of the Living Cell*

The existential reality of human beings is their non-local consciousness, which animates the body and mind. This leads us to question the brain and the mind's relationship with non-local consciousness. The brain acts as an interface between them, linking two independent systems. It has been compared to a television set that receives information in the form of electromagnetic waves from a studio and transforms them into images and sound.

Quantum biology studies the processes that take place in living beings and are based on effects characteristic of quantum mechanics. This scientific branch joins together physics, chemistry and biology. The basic quantum principles underpinning quantum biology are:

- Quantum coherence[1]

---

1 Quantum coherence refers to the ability of quantum systems to maintain synchronized phase relationships, allowing different components of the system to behave in a coordinated and unified manner.

- Quantum entanglement
- Superposition
- Quantum tunnelling

Einstein suggested the possibility of quantum processes occurring in the field of biology. In the 1930s, Bohr proposed that the principles of quantum physics could be relevant there. However, these ideas were largely ignored by the scientific community for a long time.

In his book *What Is Life?*, originally published in 1944, Schrödinger argued that biology could be grounded in quantum mechanics. Shortly before, in 1943, the German mathematician and physicist Pascual Jordan (1902–1980) had introduced the concept of quantum biology.

In the 1990s, Penrose and Hameroff jointly proposed that microtubules within neurons could act as channels for the transfer of information between local and non-local consciousness, a hypothesis first suggested to Penrose by Hameroff. These structures are part of the cytoskeleton of neuronal eukaryotic cells and are made of a protein called tubulin, composed of alpha and beta monomers in a helical arrangement. According to their theory, the brain's neuronal system forms an intricate network and consciousness obeys the rules of quantum mechanics. However, some authors opposed this theory, arguing that for it to be true, quantum coherence would have to exist in an environment at a temperature of absolute zero (-273 °C), where life is impossible.

Various scientific studies have since demonstrated that quantum phenomena occur in photosynthesis at room temperature. Photosynthesis is a chemical process that converts inorganic matter into organic matter using sunlight and carbon dioxide (CO_2), while releasing oxygen (O_2). This process is fundamental for life on our planet. In aquatic environments, it is performed by algae, cyanobacteria and various other bacteria (red, purple, green, sulphur and so on), as well as by plants; on land, it is performed by plants alone. For example, quantum coherence plays a role in the capture of solar energy by chlorophyll, a plant pigment that forms part of pigment-protein complexes. These complexes act as photosynthetic antennae that capture sunlight and transport it to chemical reaction centres. Protein complexes act as photosynthetic antennae that capture solar energy and transport it to chemical reaction centres. Some studies in quantum biology suggest that quantum coherence may help energy transfer processes to occur with remarkable efficiency, allowing light energy in photosynthetic systems to follow highly optimized pathways in extremely short timescales.

Ferritin, a protein present in almost all living organisms and involved in transporting and storing iron, is also linked to quantum phenomena. In 2021, a group of researchers from the Ningbo Institute of Material Technology and Engineering in China discovered that ferritin, which is abundant in various regions of the human brain, can transport electrons over a distance of 80 microns via quantum tunnelling. This discovery could potentially be used to treat Alzheimer's disease, for example.

Quantum processes are found in the sense of smell, DNA mutations and neuronal synapses. The migratory movements of birds are also guided by quantum phenomena. In 2021, a team of scientists from Lund University in Sweden revealed that some birds possess a quantum compass in their eyes. Far more accurate than any GPS, this 'device' is based on a protein's sensitivity to the physical processes governing the behaviour of atoms and electrons.

Quantum biology is undeniably a novel and exciting field of research, but it is still in its early stages. Therefore, it is wise to be critical of new theories.

10
UNIVERSAL QUANTUM CONSCIOUSNESS

The essential features of the implicate order are, as we have seen, that the whole universe is in some way enfolded in everything and that each thing is enfolded in the whole.
David Bohm and Basil Hiley, *The Undivided Universe*

Throughout this book, I refer to universal quantum consciousness by different terms: primordial quantum consciousness, primordial intelligence, universal quantum energy, intelligent designer and so on. I am speaking here of descending causality, the origin of everything. I believe it is worthwhile to reflect on this, as this single principle is present in all religions, which name it according to their own tradition: God, Jehovah, Allah, Brahman, Tao, Akasha and so forth.

The Catholic Church controlled most of Western European culture for many centuries. It worked hard to preserve and spread culture based on its theological religious ideology. When faced with conflicting situations, it attributed them to divine will.

Scientific development

Attempts to find rational explanations for natural phenomena began during the Renaissance. The Catholic Church tried to uphold its cultural hegemony and control by all means, even resorting to violence with the Inquisition. The friar, astronomer and philosopher Giordano Bruno (1548–1600), who proposed that the Sun was simply a star and that the universe contained infinite worlds inhabited by animals and intelligent beings, was accused of heresy and burned at the stake. Galileo Galilei (1564–1642), an astronomer and defender of Copernicus's heliocentric model, improved the design of the telescope and made observations that supported the movement of the planets around the Sun. In 1633, he was tried by the Inquisition and sentenced to house arrest, where he remained until his death.

In the 17th and 18th centuries, René Descartes and Isaac Newton, who introduced calculus, established the foundations of the scientific method. The French mathematician and astronomer Pierre-Simon Laplace (1749–1827) wrote a treatise on the universe that even impressed Napoleon. The emperor summoned him to ask how he could have written it without mentioning God even once. Laplace replied that he had no need to mention God, as reason was sufficient.

The gradual development of science has estranged humankind from the divine. Materialism, the dominant intellectual paradigm of recent centuries, has provided great material wealth, but it has also impoverished us spiritually by destroying traditional values.

Scientific answers

Scientists have always sought to answer two fundamental questions: the origin of the universe and the beginnings of life on our planet.

From the 19th century onwards, and especially in the 20th, quantum physics and advances in astronomy enabled the detailed study of the universe. Research on subatomic particles in accelerators, such as the one built by CERN near Geneva, on the France–Switzerland border, brings us very close to its origins. Today, cosmology – the branch of astronomy dealing with the origin and evolution of the universe – would be inconceivable without a quantum perspective. Supported by quantum physics, modern cosmology has shifted from a materialism linked with evolutionism to a transcendent idealism.

Thus, two main theories, materialist and creationist, explain the origin of the universe today.

Current cosmological theory is based on the Big Bang model, which posits that the universe began approximately 13.8 billion years ago. Its ongoing expansion, evidenced by the cosmic background radiation discovered in 1964, supports this hypothesis. While questions remain about what preceded this event, the modern materialist model does not propose an eternal or static universe, but rather one in constant evolution.

Many scientists have demonstrated the existence of a definite origin of the universe:

- In 1929, using the most powerful telescope in the world at the time, installed at Mount Wilson Observatory in California, the American astronomer Edwin P Hubble (1889–1953) observed that the farther galaxies were from Earth, the faster they were moving away, demonstrating that the universe is expanding. By extrapolating these results backwards, he found that everything converged on a single point: the genesis of the Big Bang, the great explosion that gave rise to the universe. The image of the cosmic background radiation captures the first light of the universe, which is why it has been described as 'the face of God'. The inflation and expansion of the universe confirm that it had a beginning.

- Einstein's theory of relativity links time, space and matter, implying the necessity of a universal origin.
- The universe will end, with some hypothetical scenarios predicting its heat death. According to the second law of thermodynamics, systems tend towards entropy and disorder. After a medium-sized star, such as the Sun, has run through its fuel (primarily hydrogen and helium), it expands into a red giant, sheds its outer layers to become a white dwarf, then gradually fades away. More massive stars can end their lives in catastrophic explosions called supernovae, potentially creating a black hole. The existence of an end presupposes a beginning.
- Achingly precise universal constants keep the universe functioning perfectly. The most minuscule variation, such as in the force of gravity, for example, would cause a cosmic cataclysm. It would be delusional to think that these constants could arise and be maintained by chance alone. Mathematical calculations of this probability result in a value that is infinitesimally small.
- The emergence of life on Earth requires the fulfilment of the anthropic principle under favourable circumstances. In cosmology, this principle states that any valid theory of the universe must account for the fact that human beings, as observers, exist. Everything is fine-tuned with painstaking

precision and nothing is left to chance. The first living being on Earth from which all others currently in existence descended was a single-celled, prokaryotic microorganism known as the last universal common ancestor (LUCA). LUCA may be directly related to cyanobacteria, which are extremely resilient in hostile environments. The continuity of life requires the capacity for self-replication, among other biological functions. This is achieved through deoxyribonucleic acid (DNA), which harbours the genetic code, and ribonucleic acid (RNA), which transports information from DNA to the ribosomes, where proteins are synthesized, and to the complex cell membrane, which acts as the cell's true brain by controlling the entry and exit of substances. These molecules and structures are so complex that their emergence by chance is inconceivable.

- Morphogenetic fields are groups of cells in the embryo that determine the development of a specific organ, containing exact instructions on its form and function. They mediate the relationship between the genotype (the set of genes that define an individual and are passed down from one generation to the next) and the phenotype (an individual's observable traits, the outcome of how genetic information is expressed). It is the cell, and not the genome, that acts as the unit of organic structure and function. In 1996, the biologist Scott F. Gilbert proposed that the previously developed concept of the morphogenetic field was a middle ground between genes and evolution. Due to their exquisite precision, morphogenetic fields, like other cosmic constants, are a manifestation of primordial consciousness, or, if you may, of the intelligent designer.

- In his phylogenetic study of species, the theoretical physicist Amit Goswami discovered a lack of gradual evolution in the known fossil record. Evolutionary biology describes a branching process from common ancestors rather than a linear progression. Birds evolved from theropod dinosaur lineages, while mammals and reptiles diverged earlier from more ancient common ancestors.
- Since the theory of evolution proposed by the naturalist Charles Darwin (1809–1882) is incomplete, we might consider that an intelligent designer, expressed as pure consciousness and independent of matter, could be the primordial force of the universe.

The universe has a purpose – an objective with a logical, intelligently evolving intention. Some interpretations of quantum mechanics have raised questions about the role of observation in physical reality, although this does not imply that human consciousness itself determines or creates reality.

There is substantial evidence for the external intervention of an intelligent designer that brought the universe into being from nothing. While the scientific method describes the origin of the universe through the Big Bang theory, entailing cosmic inflation and expansion, and attributes the emergence of life on Earth to chemical processes that produced the first life forms, both explanations still have aspects that also suggest the orchestration of an intelligent consciousness.

The study of the origin of the universe and of life on our planet brings us closer to the divine and to accepting the existence of a primordial consciousness.

Einstein asserted that the universe enjoyed perfect synchronicity because it obeyed a set of laws. Wherever there are laws, they have been established by a superior intelligence – in this case, the intelligent designer, the primordial quantum energy.

In his book *Quantum Questions,* the philosopher Ken Wilber analyses perspectives on the origin of the universe held by the founding figures of quantum physics, from Max Planck to Wolfgang Pauli. According to Pauli, we may search for the master formula of the dawn of the cosmos, but we always reach a point where the light fades to black, compelling us to accept the existence of a superior intelligence.

Theoretical physicists are actually profound mystics. The mystic seeks the primordial energy through spirituality and the scientist seeks it through science. Yet all arrive at the same conclusion: all matter originates and exists solely by virtue of a force. We must assume that behind this force is a conscious and intelligent mind. And this mind, as Planck noted, is the origin of all matter.

We have scientific proof of the existence of a universal quantum energy that created the universe and life. Our body is energy collapsed into matter; it is stardust originating from the Big Bang, set in motion by a God so benevolent and loving that He manifests within each of us as non-local consciousness. This is the understanding and forgiving God championed by the philosopher Baruch Spinoza (1632–1677), vastly different from the personal God presented as a stern judge in classical theism. Spinoza's God – an eternal, infinite and supremely perfect reality – is one of unity, harmony and love. Everything around us is divine, laid there by God for our enjoyment. The House of God is not merely a temple, but it is all nature surrounding us. Everything is the work of His hand.

Jesus Christ expressed this unequivocally (Gospel of Luke 17:21): '[. . .] the kingdom of God is not something that can be observed [. . .] because [it] is in your midst', though you do not recognize it. The entire universe is united by a force, masterminded by an intelligent consciousness. It is descending causation, the primordial energy.

11
PSYCHOLOGICAL REPERCUSSIONS OF NDES

That which we do not confront in ourselves we will meet as fate.
Carl Jung

After undergoing an NDE, patients need support and understanding from healthcare staff, their families and the people in their lives. Digesting what they have gone through can take a long time, even years, and can sometimes be difficult and distressing. Fearing the occasional lack of acceptance or understanding, patients often keep quiet about their NDE. The gulf of incomprehension can be so wide that they may even be labelled as psychopathic, leading some to receive psychiatric treatment. This is why the incidence of NDEs is believed to be higher than it appears, as many cases go unreported.

NDEs leave a profound and indelible psychological imprint. This lasts the rest of the patients' lives and affects their existential outlook, values, religious beliefs and behaviour. I have seen this directly through the periodic follow-up care I provide in outpatient clinics for patients who have had an NDE, driven by my interest in the underlying pathology of serious NDEs and their potential psychological impact.

Based on the accounts of patients who have had an NDE and my own observations, they may undergo the following changes:

- The value that the patients attach to consciousness increases, while their love for the ego subsides.
- They cultivate a growing interest in topics related to philosophy, psychology and theology.
- They become less critical and more understanding and tolerant in their interpersonal relationships. They also gain in empathy, which is the capacity to identify with others and share their feelings.
- They undergo a profound shift in their existential outlook and approach life more purposefully. They relish little details and live intensely in the present moment.
- They feel deep respect for nature.
- They clearly lose their fear of death. They are acutely aware that death bears no resemblance to what they had imagined. They possess total certainty that there is a life beyond death.

As we previously discussed (see Chapter 7), human beings fear death for several reasons:

- The transition from life to death is usually painful, distressing and desolating.
- To die is to step into the unknown.
- Death entails the loss of everything of material value acquired throughout life: family, friends, material possessions, success, fame and wealth. We come into this world with nothing and we leave it the same way.

Our powerful instinct for self-preservation makes us cling tenaciously to life. People who have experienced an NDE tend to think of birth like slipping into a body – as if it were a suit or outfit – which gradually deteriorates until the time comes to slough it off. In fact, their serene acceptance of death only crystallizes as the years go by.

Although spirituality blooms within them, these individuals often lose interest in religious affiliation. They are convinced that they came into contact with universal quantum consciousness during their NDE. Their spirituality stems from an imperative need to communicate with the primordial energy, an intimate relationship that transcends religious dogma.

When people resume their everyday lives after an NDE, all these existential and psychological changes can cause them problems, especially in managing their personal relationships. Various scientific studies have shown that an NDE can alter the dynamics of a couple's relationship, leading to a significant increase in divorce, with the rate reaching as high as 65 per cent of all cases.

However, there are positives to people experiencing NDEs. They become more intuitive. As previously explained, intuition is a manifestation of the subtle, high-frequency energy we call Supraconsciousness. Also, those who witnessed scenes of their life during their NDE remorsefully recall the harm they consciously committed against people, animals and the planet. In response, they become empathetic and kind.

12

TAPPING INTO SUPRACONSCIOUSNESS

I maintain that truth is a pathless land, and you cannot approach it by any path whatsoever, by any religion, by any sect . . . Truth, being limitless, unconditioned, unapproachable by any path whatsoever, cannot be organized; nor should any organization be formed to lead or coerce people along a particular path.

Jiddu Krishnamurti, speech at the dissolution of the Order of the Star

The identity of local neuronal consciousness has external and material origins. We call this the ego. The ego is insecure, incomplete and ever-shifting, as its stability rests on the opinions of others. It dwells on the past and projects into the future, but it bucks against the present, where Supraconsciousness reigns supreme.

> The ego is propelled by our core instincts to survive, procreate and improve our lot. These instincts drive us to seek pleasure, their only goal. Though we cannot eliminate these instincts, we can control and channel them through the exercise of our free volition.

The ego stands in direct opposition to Supraconsciousness, doing everything possible to suppress it. Our unique, authentic identity, which can collapse energy into matter, is concealed by these powerful illusions of the ego:

- **Ignorance**, which leads to a sadly widespread lack of awareness about structural reality. Entrenched powers, politicians and the media work to keep the people steeped in ignorance, as a misinformed society is easier to manipulate.
- **Selfishness**, which is encouraged by the dualistic split between observer and object. First described by Aristotle, objectivity fosters selfishness. It involves pursuing one's own benefit, pleasure or well-being excessively or exclusively, with little regard for others. It is the antithesis of altruism and generosity.
- **Attachment** to material objects, which is felt as a powerful inclination or compulsion. The voracious appetite for material possessions is linked to selfishness and a feeling of inner emptiness.
- **Fear**, which is ultimately always the terror of death. As a construct of the body and matter, the ego is entirely dependent upon them. It understands that the body is temporary and gradually decays until physical death and the ego's subsequent dissolution.

The purpose of life is the development of the self, which involves nothing less than discovering and experiencing our Supraconsciousness.

To do so, we must overcome and rein in the ego, which we could define as the 'non-self'. This is no easy task, as the ego is armed to the teeth and will battle with all its might to dominate our true self. The pre-Socratic philosopher Thales of Miletus said in this regard that the hardest thing for people to do is to know themselves and the easiest thing is to speak ill of others.

How can we allow Supraconsciousness to bloom?

An interesting perspective was shared by the French Jesuit, palaeontologist and philosopher Pierre Teilhard de Chardin (1881–1955), who was nearly excommunicated for his scientific ideas, as they ran afoul of religious dogma. In his controversial vision of human development – which led the Church to ban him from publishing his major works during his lifetime – Teilhard de Chardin argued that we start at the alpha point, in a state of impurity. We must then evolve to the omega point, reaching it once we have cleansed ourselves. This is a state of holiness and enlightenment (or Buddhahood), involving mastery of Supraconsciousness and control of the ego. When consciousness is purged of impurities, it becomes an avatar, consisting of pure energy.

We have two ways to access Supraconsciousness:

1) Unconsciously, which is how it is attained by patients who have an NDE.
2) Through meditation, a technique aimed at quieting the mental 'storm' constantly whipped up by the outside world, clearing the mind so that Supraconsciousness can arise. This involves the mastery of relaxation, breath and concentration and is strengthened by consistent practice.

Meditation: a pathway to liberation

We receive an average of 60,000 stimuli each day, 80 per cent of which are visual. Our brain must manage this relentless avalanche of information and the cognitive overload it entails. Meditation, which turns our gaze inward, is one of the most powerful ways for our consciousness, our true self, to restore balance to our lives.

Supraconsciousness reveals the value of living in the present moment, our authentic reality, where no thoughts or feelings exist. We must shed the guilt that cages us in a blemished past and quell the anxiety that catapults us into a menacing future. This is how we can control the ego, which tirelessly marshals its arsenal of illusions to come back on top.

When we attune to Supraconsciousness, we feel peace, equilibrium and joy, culminating in an expanding openness. We feel part of everything, lovingly merged with the entire universe. This path to spirituality does not run through any religious affiliation, though it is also true that for many people, sincere and profound prayer to God has the same effect.

The inner balance achieved with meditation significantly boosts our health and happiness. It alleviates stress, anxiety and tension by blocking the release of associated hormones like cortisol and catecholamines. It lets us see beyond the illusion of our material surroundings and discover true reality.

Meditation improves our physical health in many ways. For example, it lowers blood pressure and the heart rate, improves digestion, regulates glucose metabolism in diabetics, strengthens the immune system and helps to relieve symptoms in conditions such as Alzheimer's disease.

It also slows ageing through its beneficial effect on telomeres, which are located at the ends of chromosomes and protect them. Each time our cells divide, telomeres lose a small amount of DNA and gradually shorten. The biologists Elizabeth Blackburn, Carol Widney Greider and Jack Szostak won the Nobel Prize in Physiology or Medicine in 2009 for discovering telomeres and telomerase. Subsequent research, much of it involving Blackburn, has shown that poor habits such as an unhealthy diet and stress disrupt the production of telomerase, the enzyme that slows the shortening of telomeres. Meditation is highly effective in combating stress, which promotes the health of our telomeres and helps to lower the risk of suffering from age-related diseases.

Meditation promotes neuroplasticity and neurogenesis. It boosts self-esteem, optimism and vitality, sharpens concentration and decision-making, heightens intuition, clarifies the mind, steadies emotional management, strengthens the memory and facilitates rest.

It enhances both our physical and mental well-being overall and brings about morphological changes in the brain visible with fMRI. These changes occur in the volume of the hippocampus, the temporoparietal junction and the amygdala. We also observe a rise in levels of serotonin and dopamine, two of what we call the 'happiness hormones'.

Meditation increases activity in the sympathetic nervous system and blood flow to the brain, while lowering hypertension and mitigating bradypnea (abnormally slow breathing). These changes improve the function of the autonomic nervous system and benefit the cardiovascular and respiratory systems.

EEG studies have shown that certain meditative states are

associated with increased alpha-wave activity and reduced beta-wave activity, while other advanced meditation practices have also been linked to highly synchronized gamma-wave patterns. Synchronization between the two brain lobes also increases. Thus, with alpha waves predominating, an equilibrium of consciousness, peace and calm is achieved.

> Meditation requires practice and benefits from expert advice and guidance. It provides us with a way to access Supraconsciousness, or pure awareness.

Meditation is not a therapy, but a method for elevating one's state of consciousness. It leads to a singularity, which is a tear in space-time and an expansion into the pure awareness of infinity. The key is not to concentrate on an idea or force the mind, as this produces energy. One must simply let it flow.

After meditation, the mind is clear, dynamic, creative and active. Meditation improves decision-making and emotional control, creating feelings of freedom and of unity with the entire universe. In short, it enhances mental power, health and social relationships.

A society that meditates is more just, balanced, sustainable and respectful of its environment. Meditation leads us to Supraconsciousness. As such, it improves our existence, allows us to tap into our enormous potential and brings out the best in us.

At times of profound existential crisis, when the foundations of our psyche are shaken, often as a result of self-deception, the emergent glow of Supraconsciousness can seem like a beacon of hope in the darkness. For those lacking psychological support, these difficult moments can lead to suicidal thoughts or autolysis. Meditation is a conscious way to awaken to non-local consciousness, or Supraconsciousness – a form of grace or divine gift that can transform a crisis into an opportunity for growth.

It is often easier for human beings to remain in familiar patterns than to adapt to profound change.

Another conscious way to access Supraconsciousness is to embrace its vital dynamics: the archetypes of empathy, altruism, kindness, justice and, above all, love. I recommend following the advice of a highly evolved being, Mother Teresa of Calcutta (1910–1997), who was awarded the Nobel Peace Prize and canonized by the Catholic Church. The following phrase is often attributed to Mother Teresa (though its exact source is unclear): 'A life not lived for others is not a life.' This attitude must be coupled with a resolve to root out ignorance, actively seeking information from the extensive body of literature on the subject.

13
SCIENTIFIC STUDIES ON NDES

The modern preference for physical grounds of explanation leads, as already remarked, to a 'psychology without the psyche' – I mean, to the view that the psyche is nothing but a product of biochemical processes.

Carl Jung, *Modern Man in Search of a Soul*

Various scientists and healthcare professionals have studied NDEs. They have conducted this research primarily within disciplines such as medicine, psychology, neuroscience and psychiatry, using the most cutting-edge technological and scientific advancements to try to understand the phenomenon.

For example, a study conducted at NYU Grossman School of Medicine followed 567 men and women who experienced cardiac arrest during hospital stays in the United States and the United Kingdom. It was found that 20 per cent of the people who survived CPR described lucid experiences of death. These experiences occurred while the patients were apparently unconscious and on the verge of death.

This study, called AWAreness during REsuscitation (AWARE II), was conducted between May 2017 and March 2020. It was led by Dr Sam Parnia, whom we met previously in Chapter 4 (see page 34).

Gamma activity during NDEs

Other studies have found gamma activity and electrical spikes in people who have had close encounters with death, which is a sign of heightened states of consciousness.

> Gamma activity refers to neuronal oscillations in the range of 30 to 100 Hz and is associated with states of alertness, attention and sensory perception.
>
> Electrical spikes are short bursts of electrical activity that can indicate increased brain activity.

In a study published in the journal *Proceedings of the National Academy of Science* (PNAS) in 2023, researchers from the University of Michigan found evidence of a surge in brain activity for up to an hour after receiving CPR that included the activation of gamma, delta, theta, alpha and beta waves.

Five thousand cases studied

Working in a hospital in Kentucky, the renowned radiation oncologist Jeffrey Long has studied over 5,000 cases of NDEs for over 40 years. He has found that 45 per cent of those who have had an NDE describe an out-of-body experience.

According to Long's observations, these experiences occur when people are in a coma or clinically dead, providing no signs of life, yet they lucidly see, hear, feel emotions and communicate with other beings.

Long studied one detailed clinical case of a woman who suffered cardiac arrest triggered by a severe allergic reaction while under anaesthesia for an operation. After recovering, she shared her extraordinary and detailed NDE with him.

In another notable case, a woman was clinically dead for 30 minutes after drowning. Despite her state, she was able to recall vivid details of her experience, including the sensation of being pulled along by a current, encountering beings of light and feeling an overwhelming sense of peace and love.

Moody's NDE breakthrough

The name Raymond Moody (see page 37) has resonated powerfully in the field of NDEs for years. This American psychiatrist rose to fame in the 1970s for his pioneering work on them.

Moody launched his career with PhDs in Philosophy and Psychology. However, his interest in NDEs prompted him to change direction and dedicate himself to researching them. In 1975, he compiled the testimonies of many people who had had NDEs in his first book, *Life After Life.*

The book introduced the concept of NDEs to the general public, causing shockwaves at the time and rocketing Moody to fame. The book's detailed accounts presented a fascinating view of what some people experience when at death's door.

Despite his significant contributions to the study of NDEs, Moody acknowledges that his research does not provide irrefutable proof that there is another form of existence. However, his work has paved the way for future research in this field.

Today, Raymond Moody remains an influential figure in the study of NDEs. His work has inspired other researchers to explore the phenomenon and has helped to shed light on one of life's greatest mysteries: what happens after death?

A journey from materialist medicine to NDEs

Bruce Greyson, another NDE researcher we have met before (see page 37), is a prominent physician and professor of Psychiatry at the University of Virginia. For nearly half a century, he has investigated the interface between life and death.

Though trained as a materialist doctor, Greyson became intrigued by NDEs and began to investigate them, thinking he would soon find a simple physical explanation. His work, featured in various publications, has explored this fascinating question and what it can teach us about the continuation of consciousness.

Greyson is considered a leading figure in NDE research, along with the psychologist Kenneth Ring, the cardiologist Michael Sabom and others, although the earlier work of Moody and the psychiatrists Russell Noyes Jr (1934–2023) and Elisabeth Kübler-Ross (1926–2004) also stands out.

To measure the characteristics and intensity of near-death experiences, psychiatrist Bruce Greyson developed a standardized tool known as the Greyson Near-Death Experience Scale.

This scale consists of 16 items that assess different dimensions of NDEs, including cognitive, emotional, paranormal and transcendental aspects. Each item is scored from 0 to 2, so the total score can range from 0 to 32. A score of 7 or higher is generally considered indicative that the experience meets the criteria for an NDE.

In addition to such tools specifically designed for NDEs, some researchers have also studied phenomena related to the activation of kundalini energy, a concept from Indian spiritual traditions that describes intense psychological and physiological transformations.

The related questionnaires cover a wide variety of experiences, such as changes in perception, emotions, thought, sense of self, social relationships, sleep patterns and eating habits, as well as other physical and psychological aspects associated with these processes.

These instruments are primarily used for research to improve understanding of how certain profound experiences may influence our physical and mental well-being.

A bridge between science and consciousness

The renowned Dutch cardiologist and scientist Pim van Lommel has devoted much of his career to researching NDEs. After studying medicine at Utrecht University until 1971, he specialized in cardiology and worked at the 800-bed Rijnstate Teaching Hospital from 1977 to 2003.

In 1986, van Lommel began to study NDEs. Two years later, his interest led him to co-found the Merkabah Foundation, IANDS Netherlands.

Van Lommel has received many awards for his work and his popular publications have enjoyed enormous success. His book *Endless Consciousness,* first published in Dutch in November 2007, quickly became a bestseller. It has since been translated into 11 languages and sold over 400,000 copies worldwide.

Today, van Lommel continues his work as a researcher, writer and lecturer on the relationship between consciousness and the brain.

The most significant research done by van Lommel and his collaborators on NDEs was published in the medical journal *The Lancet* in 2001. This prospective study was conducted across ten Dutch hospitals and included 344 survivors of cardiac arrest.

The study focused on patients who had survived cardiac arrest and experienced an NDE. The results of the study showed that NDEs are real experiences that cannot simply be attributed to figments of the imagination, the fear of death, hallucinations, psychosis, drug use or a lack of oxygen.

Van Lommel concluded that the current materialist view of the relationship between consciousness and the brain, held by most physicians, philosophers and psychologists, is too restrictive to properly understand the NDE phenomenon. He proposed that our consciousness does not always coincide with the functioning of our brain, and that an enhanced or non-local consciousness can sometimes be experienced separately from the body. Recognizing that these phenomena are coherent with the postulates of quantum physics, van Lommel therefore believes that death could be compared to a change in the state of consciousness, as it transitions into a Supraconsciousness where space-time does not exist.

This study has been fundamental to research on NDEs and has opened new avenues for understanding human consciousness. Though it does not focus on any specific clinical case, it provides a valuable perspective on NDEs in general.

14
HOW I DISCOVERED DR MANUEL SANS SEGARRA

–Juan Carlos Cebrián Barrientos

The real voyage of discovery consists not in seeking new landscapes, but in having new eyes.
Marcel Proust

I had never been a particularly spiritual person and I never used to be drawn to theories about what happens after death . . . until April 2023. I only write this chapter to explain how I met Dr Manuel Sans Segarra and how I became interested in these topics. From there, each reader may draw their own conclusions.

As a child, I shared a powerful bond with my maternal grandparents, Isabel and Francisco. My mother's twin brother, Andrés, died in an accident at work one week after I was born, in May 1977. This meant that I spent a great deal of time with my grandparents during my childhood and teen years. In just seven days, their grandson was born and they lost their son.

I slept under the same roof as my parents, Pepe and Paqui, but during the day and at the weekend, I spent almost all my time at my grandparents' house. The reason I'm telling you this is because there is a connection between the moment I met Dr Sans Segarra and my relationship with my grandparents. Now more than ever, I feel that they are with me somehow and are guiding me. It is difficult to put into words.

I studied Journalism and have launched several entrepreneurial projects, none of them related to spirituality or the afterlife. In fact, whenever anyone mentioned anything related to those topics, I didn't pay it much mind, as I couldn't see any logical reason for it.

My grandmother passed away in 2008 and my grandfather in 2012. In the final months of my grandmother's life, we looked for a caregiver, and we found one by chance through an online advert. Manana arrived, a highly cultured Georgian woman who was nearly fifty at the time. She had been a university music professor, but upon arriving in Spain, she had to work by caring for the elderly.

As we had cultivated a wonderful relationship with Manana, she stayed with our family after my grandmother died and helped us to care for my grandfather in his final years.

Manana's family remained in Georgia and I went to visit them twice. On my first trip, her mother, Eliko, who had been a textile engineer, made a special impression on me. Even though she was not fluent in either Spanish or English, Eliko made herself understood and radiated kindness and love from every pore. I saw her twice in my life.

The first time we met was just a few months after my grandmother Isabel had died, and I was immediately flooded with a powerful feeling of tenderness and love for Eliko. It was strange and unsettling, because I had never seen her before and we only talked for a few hours, despite having no language in common.

Eliko was about to turn one hundred in September 2023. It was going to be a major event and Manana was obviously going to be there to celebrate. However, on Sunday, 23 April 2023 (Saint George's Day), Eliko passed away at her home in Tbilisi, surrounded by her grandchildren and friends. Saint George's Day is one of my favourite days of the year, and it is also an extraordinary day to visit Barcelona, as it is known there as the Day of Books and Roses.

All that week, I wanted to find out more about Eliko and came across a social media post written in Georgian. The Georgian alphabet is an *abjad* alphabet, completely unrelated to the Latin one. To help you to understand what this means, I will write Eliko's full name in both the Latin and Georgian alphabets, then explain what happened.

In the Latin alphabet, her name is spelled Elene Sepiashvili.

In the Georgian alphabet, it is: ელენე სეფიაშვილი

I only know about three words in Georgian and I cannot read the alphabet to save my life. However, that social media post caught my eye, and since I knew it was about Eliko, given how recently she had passed away, I copied the text above, and not any other text on the page, and ran a Google search to find out more about her.

This article appeared on the screen:

*http://silkmuseumblog.ge/2020/11/20/*ქართული-აბრეშუმი-*1950-90*-იან-წლე/.

Fortunately, the website had an English version too:

http://silkmuseumblog.ge/en/2020/11/19/georgian-silk-in-1950-90s/

ელენე სეფიაშვილი

თბილისის აბრეშუმის საქსოვი ფაბრიკის ინჟინერ-ტექნოლოგი

დაიბადა 1923 წლის 1 სექტემბერს წითელწყაროს რაიონის სოფელ არბოშიკში. საშუალო სკოლის დამთავრების შემდეგ სწავლობდა თბილისის ი. გოგებაშვილის სახელობის ქართულ პედაგოგიურ სასწავლებელში. 1941-42 წლებში თბილისის სტალინის სახელობის სახელმწიფო უნივერსიტეტის ფიზიკა-მათემატიკის ფაკულტეტზე, რომელიც არ დაუმთავრებია. 1946 წელს გადავიდა მოსკოვის საფეიქრო ინსტიტუტის თბილისის ფილიალის ტექნოლოგიური ფაკულტეტის III კურსზე, რომელიც დაამთავრა 1949 წელს. ამავე წელს დაიცვა სადიპლომო ნაშრომი ქ. ივანოვოში და მიიღო ინჟინერ-ტექნოლოგის წოდება ქსოვის სპეციალობით. 1947 წლიდან ელენე სეფიაშვილი პარალელურად მუშაობდა თბილისის აბრეშუმის საქსოვ ფაბრიკაში წუნმდებელ-კონტროლიორის თანამდებობაზე, მოგვიანებით მოსამზადებელი საამქროს, შემდეგ კი საქსოვი და მოსამზადებელი წარმოების უფროსად. დაინტერესებული იმით, თუ როგორ მუშავდებოდა ახალი ქსოვილები ელენე სეფიაშვილი დაინიშნა ქსოვილის ასორტიმენტის დამმუშავებელი (დესინატორების) ჯგუფის უფროსად, რომლის შემქმნელი იყო ვასილ ივანეს ძე პეროვი.

In the photo, Elene Sepiashvili is the woman on the left. Once again, I was looking at an article written in the Georgian alphabet. Unable to understand what it said, I used an online translation tool and this is what appeared:

Elene Sepiashvili

Engineer-technologist at the Tbilisi Silk Weaving Factory

Elene Sepiashvili was born on 1 September 1923 in the village of Arboshiki, in Tetritskaro District. After graduating from secondary school, she studied in Tbilisi at Iakob Gogebashvili Pedagogical School. During 1941–42, she attended the Faculty of Physics and Mathematics at Tbilisi State University, which bore Stalin's name at the time, though she did not graduate. In 1946, she transferred to the third year of studies at the Faculty of Technology in Tbilisi, part of the Moscow Textile Institute, and graduated in 1949. That same year, she defended her thesis in Ivanovo and was certified as a textile engineer. Starting in 1947, Sepiashvili also worked at the Tbilisi Silk Weaving Factory, first as a quality controller, then as head of the preparatory workshop, and finally as head of weaving and preparatory production. Interested in how new textiles were developed, she was appointed head of the group of processors (designers) for fabric assortment, a group founded by Vasili Perov.

Unexpectedly, the supervising authorities wanted to produce as many beautiful textiles as possible during the war years. The workshop also produced warp and weft threads, which were delivered to the Weaving Department. The Tbilisi Silk Weaving Factory developed gradually. The joint group, headed by Sepiashvili, put new machines into production. Experimental looms were also allocated to create new fabric assortments.

The textiles created by Elene Sepiashvili were produced at the factories in Tbilisi and Kutaisi, as well as in Moldova. The products were launched in Georgia and the technical data was sent to other countries whenever there was interest. Sepiashvili often successfully presented new assortments at the annual All-Union Council held in Moscow. A natural silk fabric she created for dresses, called Tsitsinatela ('Firefly'), received a silver medal at the All-Union Exhibition in Moscow. Another textile she created, Vardzia, enjoyed high market demand.

Sepiashvili made great progress in her career and was awarded the distinction of Honoured Engineer of the Republic. She felt inspired to be a successful engineer and to represent her country. During a period of production cutbacks, silk production came to a halt. However, Sepiashvili began to lead a small weaving production outfit at the Tbilisi Polytechnical Institute, where she worked for several years and created various samples of consumer fabrics.

I shared this article with my entire family and with Manana, who had been unaware that it had been written about her mother. I knew that Eliko, as she was called, had been a textile engineer, but not that she had studied Physics and Mathematics. One can only imagine what this meant in Soviet society in the 1940s.

From this point, things begin to happen. I leave it to you to interpret whether the events in question were random or charged with meaning.

And I am absolutely certain that none of them would have come to pass if Jordi Juez, my brother-in-law and a fundamentally decent person, had not encouraged me to start playing padel again in late 2022.

On the padel court I made a new friend, Juan Antonio Fernández, a highly spiritually evolved person with penetrating intuition. He introduced me to Antonio, a man of great sensitivity, whom I also consider a friend. Antonio told me about Dr Andreu Gabarrós.

And thus flowed forth a stream of circumstances that I am about to explain. I feel a powerful bond with these individuals and they are part of this book's story, even though we come from totally different worlds.

The day I heard Dr Sans Segarra for the first time

On Sunday, 30 April 2023, one week after Eliko's death, I woke up intending to search for a video on YouTube made by Dr Andreu Gabarrós, the head of the Neurosurgery Department at Bellvitge University Hospital. A specialist in brain surgery, renowned in Spain and abroad, Dr Gabarrós has over 25 years of professional experience in clinical practice and research.

I wanted to find a video about the project 'Symphony of Heroes' that Gabarrós had created. This project is a tribute to the determined efforts of patients who have undergone brain tumour surgery. Gabarrós selected nine patients who were musicians, composed a song for each of them and adapted it to the instrument the patient played. He then recorded an album featuring them all.

Instead of showing me this video, YouTube presented me with one featuring Dr Manuel Sans Segarra, whom I had never heard of before. Under normal circumstances, I would have continued searching for what I wanted, but that day, I stopped and watched the whole thing.

I was deeply moved. I cried. I had never been interested in NDEs before, but that video awakened my consciousness. I was absolutely certain that I wanted to meet him and it became one of my top goals for the week.

Here I have to thank his highly intuitive wife, Magaly. Even though she had never met me before, she played a key role in how I was able to meet her husband and why this book was written, setting in motion a train of consequences that I believe is awakening consciousness across the world.

She insisted that he speak with me after my first call and I ended up talking with him. As I write this now, I know that this was the only time that it would have been possible. Before that day, I was not ready and did not have the time necessary to help him. And I know that it would have been practically impossible to reach him afterwards, due to the thousands of messages he now receives each day.

And here, I am going to explain something that, as I said earlier, you may interpret as you see fit. At that very moment, I felt it was Eliko's way of thanking me for sharing her story with my family. That woman, that physicist, mathematician and textile engineer who lost the status she had built over so many years with the dissolution of the USSR in the late 1980s, and whom I had only seen twice in my life, was reaching out to me. That morning, I felt completely connected to Eliko, just as I had been with my grandparents when I was a child and a teen.

In fact, when I look now at the photo of Eliko in the textile factory that I found in the online article, her gaze reminds me of my grandmother Isabel's. As I mentioned before, it is deeply unsettling. It leaves me speechless. As I write this, I simply let myself get swept away by whatever thoughts come to mind. After reading this book, I am beginning to understand that this is related to certain moments of interrelation. It is not always like this in my everyday life.

The lives of Eliko and my grandmother Isabel roughly overlapped. Eliko was born on 1 September 1923 and my grandmother Isabel was born on either 18 March or 18 May 1918. I should explain this confusion, as it is also rather interesting. I was born on 18 May 1977 and she always said she had also been born on that same day in 1918, but her birth certificate reads 'March'.

In any case, my grandmother never celebrated her birthday again after the death of her son, Andrés. And so, throughout my childhood and teenage years, we only ever really celebrated my birthday on 18 May.

Eliko and my grandmother Isabel were alive at the same time for many years, despite being thousands of kilometres apart. According to the laws of physics, there was only the remotest probability that they might ever cross paths. In fact, they never did meet. And yet, as I write this, I feel that everything is connected.

My discovery of NDEs

A few months after I spoke with Dr Sans Segarra, first by telephone and later when I visited him at his office in the Barcelona College of Physicians, I also had the opportunity to meet and interview Dr Andreu Gabarrós in person. Both had worked at the same hospital and had crossed paths. I did not explain this whole story to Gabarrós, though we did speak about life and death and all the work and musical projects he is carrying out with his patients. Once again, it was as if everything was somehow connected.

Jesús Alonso Gallo and Tessa Romero, the two people I describe in later chapters of this book who have reported NDEs, also came into my life through incredible coincidences. I was subscribed to Jesús's newsletter. He is a businessman and investor who I initially assumed had nothing to do with transcendental issues. That was, until one day I read something he wrote discussing quantum biology, without him even realizing it, as his newsletter is about life and entrepreneurship. I had been reading about that same subject elsewhere, so I wrote to him. And he wrote back.

We arranged to talk and he explained that he had an NDE in 2008, which I will detail later. Furthermore, since I met Jesús, a tapestry of interlaced events has been woven from a golden thread. When I think about each person who has led me to him and what has happened since, all are deeply sensitive and have somehow managed to reach a higher state of consciousness at some point in their lives.

Tessa's case is also remarkable. I was driving in my car, listening to music and podcasts related to entrepreneurship, which is what I usually listen to while driving. One business podcast ended . . . and Tessa's voice came on, explaining how she had been clinically dead for 24 minutes. Since my consciousness had been awakened by then, I listened to the entire two-hour interview. I was astounded. I contacted her a few months later and the chapter on her NDE (see Chapter 15) is the result.

One might think that all of this is merely a string of coincidences. And maybe it is. But, as Dr Sans Segarra says, this book is addressed to curious readers, who are eager to know more and who obviously question everything they read. From doubt, great thoughts emerge and profound growth takes place.

15
TESSA ROMERO'S NDE

–Juan Carlos Cebrián Barrientos

The writer and speaker Tessa Romero rarely grants interviews. Given that several very interesting podcasts and interviews with her can be found online, we decided to take a different approach to writing this chapter. First, we drew on that publicly available material, then we spoke with her to flesh it out and add the final touches.

Tessa has published three books, one of which is a bestseller entitled *24 Minutes on the Other Side: Living Without Fear of Death.* First published in Spanish in 2018 and read by millions of people, this book deals with her NDE and how she learned to live without the fear of dying. In addition to being a writer, Tessa is also a sociologist and journalist. She has travelled the world doing humanitarian work with non-governmental organizations.

She explains why she doesn't usually give interviews: 'What I like is writing, so I can help people'. She prefers to spend her time answering emails and talking to people via social media.

Tessa comes from Andalusia, but she has a special connection to Latin America. Her daughters were born in Costa Rica and she has been to the region several times on humanitarian missions since she was twenty. She values the warmth and friendliness of the people who live in those countries.

Her experience started with the diagnosis of a rare and incurable illness. Doctors told her there was no treatment and that her body was shutting down. They gave her a week to live, or ten days at the most.

Tessa was terrified. She decided not to tell her family that she was going to die. She could not bear the idea of causing them any more pain, especially her mother, who had already suffered several losses. Instead, she wrote a letter to each of her loved ones, saying goodbye for when the time came. She also arranged custody for her daughters and formalized everything in writing with a notary.

Throughout that time, Tessa kept silent about her diagnosis. Even though her body began to show signs of illness, she insisted that she was fine. She even attended a family wedding, where people commented on how thin and ill she looked.

A few days later, while picking her daughters up from school, Tessa began to feel the symptoms of a heart attack. She knew she was reaching the end, but she decided not to go to the doctor. Instead, she chose to surrender to whatever was meant to happen.

Twenty-four minutes on the other side

While waiting for her daughters outside the car, she felt the heart attack tightening its grip. Suddenly, a woman with two young children rushed to her aid and took her to a small clinic right across the street. Although Tessa cannot share the whole story here, out of respect for those who read her book, she thanks that woman for saving her.

Facing death at a young age, Tessa was drowning in immense grief. She yearned for it all to be over as soon as possible. In the midst of her anguish, she found herself on a gurney, disconnected from her body, yet still bound to it. She no longer felt any physical or emotional pain and was blissfully comfortable.

At some point she realized that the body on the gurney was her own. As the medical team tried to resuscitate her, there were moments when it felt as if they were pulling her downwards. In one of these moments, she saw and recognized herself. Though she was not aware that she was dying, she pleaded not to be roused and to be left to drift off, because the pain had finally lifted.

Then something tugged on her and she found herself before a being that received her. There was a blinding light. Still unaware that she was dying, she felt irresistibly drawn to its radiance. Since her experience in 2007, Tessa has researched the subject, collecting testimonies. According to her findings, everyone experiences the same thing upon seeing the white light: they feel drawn to it, as if returning home after a long journey.

Tessa believes that our souls know how to be born and how to die.

According to her, no spirit remains trapped anywhere, weeping or suffering. All souls migrate to the same place; all are ultimately reunited in the same realm.

During her experience, Tessa encountered beings of light and the one she considers her spiritual guide. Even though she had never heard of such concepts before, she was moved nearly to tears, because it all felt so utterly familiar. She understood that this entire process was about going home. Although she could not explain this belief, she could feel it.

Tessa does not mention seeing a tunnel, which is commonly associated with NDEs. Instead of a tunnel, she describes a dazzling white light. Sometimes, she says, the light can be circular, enveloping everything, or it can appear as a huge horizontal doorway.

When she saw the light, she felt a sense of well-being and of being home. Although she was not conscious of dying, she knew that people were waiting for her.

In the midst of the radiance, Tessa saw someone she knew, who had died many years earlier. This person, who was surprised to see her there, prevented her from crossing into the light. Tessa was angry with her father, who had died when she was 24, for not being there to welcome her. However, she later understood that if she had seen her father, she would have gone with him, and that was not meant to happen.

This figure standing in the light was telling her, 'No, no. Not yet.' Tessa did not understand what it meant, but she knew that she wanted to move into the light. She felt that the light was her home – the home of her soul.

At that moment, she saw beings of light behind the figure.

These beings can take different forms for each person, depending on what they find most comforting. In Tessa's case, all the beings of light were identical, except for her spiritual guide, which had slightly distinct features.

In several places at once

These beings of light took her on a journey through the universe. Although it is difficult for her to describe, Tessa remembers seeing the Sun and the Moon and recalls travelling faster than the speed of light. At one point, she saw two gorgeous lights and the beings of light indicated that she focus on them. Years later, she saw a photo online that reminded her of what she had seen during her NDE.

Tessa realized that she was dead when the beings of light brought her back to Earth. But the Earth she saw was different. She found herself in a green valley, with a beautiful lake below, surrounded by nature in its purest state. She could see the animals and feel the heartbeat of all living beings, even the air itself. In that moment, she understood that she was on Earth, but in a form different from the one she knew.

During her experience, Tessa was in several places at once. She was in the kitchen with her mother, who felt her presence and thought it was the spirit of her dead father. She was also in the car with her sister, trying to speak to her, but her sister could not hear her.

By the lake, she saw someone who was about to take their own life. Her spiritual guide told her that she could stop them if she wanted to. Without fully understanding what that meant, Tessa ran towards the person and tried to embrace them to prevent them from jumping.

The person continued to cry and Tessa felt a profound sadness for the first time during her experience.

The beings of light told her that she could help to prevent the person's death, but also that there was very little time. However, before she could act, she had to return to her physical body. Reintegrating into it required tremendous effort, but ultimately she succeeded.

When Tessa opened her eyes, the doctors and everyone present were amazed. They could not believe she was still alive, as she had been clinically dead. Tessa came to just as the ambulance arrived. The doctors were at a total loss to explain what was happening.

Tessa told them everything she had experienced during her NDE, including the conversations she had overheard and what the doctors had done. She described the tiniest details with such excruciating precision that they were left dumbstruck. They were bewildered that she could know all that since she had been clinically dead.

The first words Tessa uttered when she opened her eyes in the intensive care unit were: 'We are all one, and everything is one.' In that moment, she felt united with every person there – the doctors, the orderlies, the nurses and the patients. She felt that we are all one and that she was a part of everyone.

Why we live and why we die

Tessa has reflected at length on how human beings have spent thousands of years asking why we are here. After all she had experienced, she concluded that life is actually quite simple. She learned that we are here to serve others, to love, to live and to be happy. She now understands that most of our fears only exist in our minds and that society benefits when we live in a state of fear.

She also thinks that our culture regards death with terror. However, after her experience, she realized that we have nothing to fear by dying. Her story is a testament to human resilience and an inspiration to those facing similar challenges. Her experience tells us that, while death can be a time of sadness and grief for the loved ones left behind, it can also be an experience of returning home – of going back to the place from which our souls departed before coming to Earth.

After her brush with death, she now sees it as a loving being that offers a hand so we do not have to cross to the other side alone. For her, dying is like traversing a wondrous bridge. Death is love driving a horse-drawn carriage to carry us back home.

Tessa realized that we are all one and that we are all part of something much greater than ourselves. Although she did not believe in God before her experience, she now wholly trusts in a higher power. She calls it 'God' or 'the Universe', but what is important is not its name, but only that it exists and that we are all part of it.

She learned that when a loved one has decided that they want to go, we must let them die in peace. Instead of surrendering to weeping and sorrow, we should show them our love and bid them a serene farewell.

Tessa also believes that many people on the threshold of death can impart profound life lessons and face their transition with dignity. She learned a great deal from the terminally ill and dying when working to care for and support them.

According to Tessa, our loved ones who have passed away are still with us. Even though we cannot see them, we can feel their presence and receive signs from them. Tessa believes that if we are open and willing, we can perceive these signs and understand that our loved ones are here right now.

Death by suicide and the death of animals

Tessa Romero has worked with people who have tried to take their own lives since 2007. Based on her accumulated experience, she concludes that people who die by suicide never found their way in life. However, that does not mean that their death is tragic or shameful. On the contrary, Tessa believes that every life has a purpose and meaning, including those lives ended deliberately by suicide.

Tessa is critical about how society and some religions treat suicide as taboo. She argues that this only stigmatizes those who have died by suicide, along with their bereaved families. Instead of fostering shame, Tessa believes we should talk openly about suicide and offer robust support to those struggling with suicidal thoughts.

Tessa has found a new purpose in life after her transformative experience, and now she strives to help people through her work and writing. Her books, such as the aforementioned *24 Minutes on the Other Side* and *Spirits in the Bonfire: Hope and Grief in Illness,* detail her experience and offer comfort and guidance to those dealing with the loss of a loved one or struggling with suicidal thoughts. In 2023, she published *Los animales también van al cielo* (Animals Go to Heaven Too, not yet translated into English), which addresses the rarely discussed topic of what happens to animals, including our pets, when they pass away. She plans to continue writing books that explore not only grief and hope, but also the magical bond that unites people and the mystery of life.

According to Tessa, love is indestructible, as it is the most powerful force in the universe. Long after our loved ones have moved on, their love for us endures. Over time, that love becomes woven into the fabric of our lives, mending the tears rent by their absence.

16
JESÚS ALONSO GALLO'S NDE

–Juan Carlos Cebrián Barrientos

Jesús Alonso Gallo has left a deep mark on the business world over the past 40 years. His career is a testament to versatility and tenacity, having held leadership roles in marketing and sales at various companies and having founded four of his own. Three of these companies were later purchased by major players. For example, DROsoft was acquired by the US multinational Electronic Arts, and Restaurantes.com was sold to the Michelin Group.

Today, Jesús is not only a serial investor, but also an educator and mentor who shares his experience and knowledge with the next generation of entrepreneurs. He works with several different public and private institutions, including the Complutense University of Madrid, the School of Industrial Organization (EOI), the EAE Business School and the Institute of Positive Thinking (IPP), where he is involved in the Master's programme for Entrepreneurship.

Jesús had two clear goals from a young age: to own a red, two-door BMW and to become a millionaire before the age of 30. To achieve these goals, when he was 18 he got a job as an insurance salesman to pay his way through university. He always considered himself a student regardless of his position, absorbing knowledge and adding value to the companies he worked for.

His career took a turn when he learned that some friends were planning to launch a video game startup company. When he was just 23, he joined the project as its sales director. The company, Dinamic Software, found success during what became known as the golden age of Spanish software. Jesús continued to climb the ladder, holding leadership positions at the video game companies Dinamic Multimedia, DROsoft and FX Interactive. Dinamic won the 1987 Best Company Award, which landed Jesús and his team on the cover of *El País* newspaper's Sunday magazine supplement *El País Semanal,* under the headline 'Computer Geniuses'.

After selling his stake in FX Interactive, Jesús decided to leave the video game industry. With money in his pocket, he embarked on his fourth venture, this time in the food and restaurant sector, one of his great passions. He created a free online reservation platform, Restaurantes.com, which became the top one in Spain. The Michelin Group later acquired the company, fulfilling the prediction that Jesús had once written on a Post-it note stuck to his refrigerator door. And this pivotal achievement occurred only after an NDE that would change his life forever.

A surgical error that changed his life

Let us set the scene in the days leading up to Jesús's NDE. Accustomed to undergoing operations for granulomas on his vocal cords, he had another such procedure scheduled for late October 2008. For him, these had become routine: just one more trip to the operating theatre.

At the hospital, he was assigned room number 13, which made him and his wife, Esther, feel somewhat uneasy. However, there was a palpable shift in the atmosphere. Esther's tenderness made a notable difference, forging a bond that transcended the cold, clinical environment.

News that the operation was delayed gave them an unexpected gift of time. Jesús and Esther made the most of it by stealing into the bathroom for a moment of passion that lifted them beyond their solemn surroundings. Their intimate union helped them to bridge the chasm between their preoperative anxiety and the promise of recovery.

In the operating theatre, Jesús was unable to repress his entrepreneurial spirit and chatted with the medical staff about the venture that consumed his energy at the time, Restaurantes.com. However, his attention was diverted by a laser scalpel, unwrapped from its packaging as if it were a household item. The critical moment came when the laser scalpel burned Jesús's glottis shut. The staff's desperate, scrambling attempts to save him were seared into his memory before he sank into a drug-induced sleep.

When Jesús woke up abruptly in the ICU, he felt disconnected from reality. Restrained and struggling to communicate, he reeled in a frenzied mayhem of blood and needles. Esther entered, dressed like an astronaut to access the sterile area. The sheer surrealism of it nearly punctured his shock, opening a view of raw terror etched onto her face.

Each day in the ICU was a test of endurance. He suffered a perforated lung, a subcutaneous oedema that puffed him up like the Michelin Man and temporary loss of sight in one eye.

Each word Jesús expressed, written in pen on paper, was a triumph over adversity. Those 13 days and nights in the ICU were not simply a physical struggle, but they also plunged him into the complexity of the human experience of looming death. Each heartbeat was an act of resistance, each breath a victory on his road to recovery.

The moment when Jesús left his body

While Jesús was receiving intensive medical care, severe vascular damage caused significant bleeding from his arm. A cardiologist, who happened to be visiting one of his own patients, burst onto the scene. He shouted at the nurses, demanding the medication needed to stop the haemorrhage.

A nurse diligently prepared two units of the medication, whose name Jesús can barely recall. He felt the cold liquid course through his bloodstream. An organ was failing, but he could not tell if it was his heart or his brain. Everything went blurry and he passed out. Then Jesús found himself outside his body, up in the air and well off the floor.

The cardiologist began suturing the hole from which the blood was pouring. Upon waking, Jesús was surprised to find that he was still alive. This episode would repeat several times, each one carrying him back to the experience of being outside his own body, observing from the periphery as he lost consciousness.

He was admitted to the hospital in late October 2008 and remained in the ICU for nearly two weeks. After a few more days in a standard ward, he was finally discharged. The critical recovery period delayed the launch of Restaurantes.com until the following May. However, there was far more to Jesús's traumatic stay in hospital, and following his discharge, than this.

Before the anaesthesia took full effect, he reacted instinctively and the medical staff had to restrain his limbs. In the midst of it all, Jesús thought about accepting death. 'In those moments,' he recalls, 'all my worries about my family evaporated, leaving me feeling calm and accepting.' Floating outside his body, he watched the medical staff's efforts from a unique vantage point, as if waiting for the next chapter of his existence to begin.

Jesús then entered the first stage of his NDE. As mentioned earlier, these stages are deduced from over 10,000 well-documented clinical cases.

In the first stage, everything felt hyperreal. Individuals who have had an NDE describe this as more real than everyday life. He remembers thinking only of himself during this period, submerged in deep introspection.

He also passed through the second stage, which involves an out-of-body experience. In this stage, people feel they have left their physical body and observe it and the events around them from an outside perspective.

Jesús floated in the air, watching the frantic doctors from above. He felt nothing like the physical sensation of standing on the floor.

However, he did not experience the third stage, when some patients report hearing others' thoughts or knowing about events happening elsewhere at the same time. He acknowledges that he did not reach that point, nor did he perceive anything unusual, though he would have liked to.

Jesús felt profound peace during this time, which ended up transforming his perspective on life. 'I cannot quite remember if I encountered other people or whether I had any conversations with my grandfather, Gabriel. He was an entrepreneurial man who had a huge impact on my childhood. I had a tremendous bond with him.'

Facing a future without speech

After his recovery, Jesús grappled with the prospect that he might never be able to speak again. Supported by a team of psychologists and psychiatrists, he faced the challenge of building a new life in the wake of his experience.

'When the doctors told me I might never speak again for the rest of my life, I wrote in a notebook, replying that I made my living by talking,' Jesús says. 'They explained that technology had advanced and that I could speak through a voice synthesizer on the computer. They assured me that I would be happy, even if I couldn't talk. Thanks to our efforts to prepare for that, I left the hospital having accepted that my life would be different. The miracle is that what I do now is connected to that experience. Now I write every day, which I started in the ICU when I couldn't speak, even though I can speak now. This allows me to fulfil my mission, which is to help others to launch businesses, to invest, to be happy and to realize that life is a gift.

'To regain my voice, I changed medical teams after the first operation. I went to the head of otorhinolaryngology at La Paz Hospital in Madrid. I underwent another procedure and they fixed the problem. With the help of a speech therapist, I was surprised how much of my voice returned.

'In my life, I have had experiences that defy explanation; some things go beyond mere coincidence. For example, after my divorce, I became reunited with, then married Esther, the girl I fell in love with when I was a boy. Every seeming hardship, like my divorce, turned out to be a blessing from God, even though I was an atheist until my NDE. Then, in the middle of all these meaningful events, my brother beat cancer that was supposed to be incurable. He also had children, overturning every medical prognosis, as he had been told that he was totally infertile.'

For Jesús, the lesson is perfectly clear: 'The conclusion I've reached is that I'm privileged, and that everything has happened to me for a reason. From being an atheist, to agnostic, to becoming certain that God exists, I feel His presence every day. In the *Iliad* of my life, God watched over my death and resurrection. My experience and my brother's experience seem scientifically inexplicable, which only strengthens my belief in the divine.'

A life full of meaningful coincidences

When Jesús was 40, he married for the second time. This time his bride was Esther, whom he had first fallen for as a boy. But before their reunion, his brother was diagnosed with terminal bone cancer.

His brother underwent extensive surgery, but it offered little hope. Though weakened by the illness, he started to look after Jesús and his first wife's infant son while they were at work.

This was a profoundly meaningful time for Jesús's brother, who found the strength to care for and feed his nephew. As he did so, he gradually began to heal. Their mother, a devoutly religious woman, considered it a miracle from God, the result of many people's prayers. The oncologists were unable to provide a scientific explanation for his recovery.

Jesús trusts in his feelings about God and is convinced that He exists. He does not express a preference for any particular religion, but he experiences God's presence in everyday aspects of his life. This is reflected in the peace he feels when he recognizes it in music, in the eyes of children, in animals and in nature.

A magnificent conversationalist with a brilliant mind, Jesús continues to receive a constant stream of life-affirming synchronicities. After suffering a heart attack in January 2023, he has written a daily email about life, entrepreneurship and investing. If you would like to receive it for free, just sign up on his website, www.jesusalonsogallo.com. It does not matter whether or not you are a business owner or an entrepreneur, as you will have the chance to read the thoughts of one of the most interesting people I have ever met.

17

DR JOSÉ MORALES DEL RÍO'S NDE

–Juan Carlos Cebrián Barrientos

This final chapter devoted to a real NDE posed a special challenge. We had finished writing the rest of the book months earlier and only lacked the personal testimony of a healthcare professional or a physician who had undergone an NDE.

We found several such individuals who fulfilled our criteria, yet for various reasons, they ultimately decided not to share their stories with the world. Only time will tell if they eventually change their minds.

It is not easy to share an NDE, as you can never predict how colleagues, friends and family members will react.

Upon learning that these accounts would remain private, I told Dr Sans Segarra and our publishers that if we did not find the testimony we sought in time, we would have to release the book as it stood.

A great deal has happened in the three months between writing the previous chapter and this one. Public interest in Supraconsciousness and scientific evidence for life after life has surged, as suggested by the rising engagement on Dr Sans Segarra's YouTube channel and social media profiles.

Healthcare professional or physician?

We believed it was essential to include a personal account of an NDE from someone whose professional standing provided unassailable credibility and authority.

Earlier in this book, Dr Sans Segarra explains his first-hand experiences with patients. One of these patients was a hospital colleague who survived a serious traffic accident.

In fact, many prestigious physicians have had NDEs. One of the best-known and most respected is the American neurosurgeon Dr Eben Alexander, a former professor at Harvard Medical School and the author of the bestseller *Proof of Heaven*. In 2008, Alexander had an NDE while in a coma, battling bacterial meningitis.

During his NDE, Alexander reported travelling to a celestial realm filled with beauty and peace. He described seeing a brilliant light, angelic beings and fields of multicoloured flowers. He also felt profoundly united with universal love and wisdom.

After he came out of the coma, Alexander felt completely transformed. The experience reshaped his worldview and convinced him of the existence of an afterlife.

Until that moment, Alexander had been opposed to considering NDEs as a phenomenon with scientific merit, dismissing them as hallucinations or even irrelevant nonsense. In *Proof of Heaven* (2012), he recounts, 'As a doctor of Medicine, with a long professional career at prestigious institutions like Duke and Harvard, I was the perfect sceptic. I was the kind of guy who, if you told me about your NDE or a visit from your deceased aunt assuring you everything was fine, would have looked at you and said, compassionately but firmly, that it was a fantasy.'

The case of Dr Morales del Río

On the morning of Wednesday, 20 March 2024, I was preparing the first videos for Dr Sans Segarra's YouTube channel with his production team.

I told Dr Sans Segarra that I still had not found anyone with the necessary credibility who was willing to explain their NDE for the final chapter. A few weeks before, I had received a message from a practising family doctor, but she ultimately declined to share her story. I understood, of course, but I still wanted to find the ideal testimony to complete a book whose primary purpose is to help people. Now, with Dr José Morales del Río's NDE, I feel that we have done just that.

That evening, I was back on YouTube and discovered four Spanish doctors explaining their NDEs online, in varying levels of detail. I immediately wrote to three of the four. One of them, Dr José Morales del Río, replied in just a few minutes.

Another doctor replied a couple of days later, though we have not yet moved ahead with her story. Perhaps it can be part of another book.

I spoke with José on the phone for almost an hour that night. Once again, there were coincidences that you may interpret as you see fit.

José was the guest in the video that immediately preceded the one with Dr Sans Segarra on the YouTube channel where I first saw him on Sunday, 30 April 2023. In fact, José told me he already knew of and had met Dr Sans Segarra, and that he had shared this with the journalist who interviewed him, Maria Martí, who runs the YouTube channel *Mientras Viva* (While I Live).

While writing this chapter, I spoke with Maria, who said that she also previously knew about Dr Sans Segarra, even before the interview, having seen him at a conference in the Catalan town of Calella.

Devoted to family medicine

Dr José Morales del Río was born and raised in a working-class neighbourhood in the Catalan city of Lleida in 1960. Despite his family's modest social standing, his childhood was a happy one. A gentle and introverted homebody, he cultivated a deep passion for reading and music and pursued both as best as his environment and the times would allow.

When preparing for university, he was at first drawn to Psychology at the University of Barcelona, as he was fascinated by the complexity of human behaviour. However, circumstances led him down a different path. In 1977, the first intake for a degree in Medicine began in Lleida. Although it had not been his first choice, the opportunity would allow him to study while living at home, offsetting the significant expense of living in Barcelona and thereby easing the financial burden on his family.

Once immersed in his studies, his interest turned towards neurology and psychiatry, disciplines that explored the mysteries of the brain and the human mind. Yet fate ultimately steered him towards family medicine, which would enable him to provide his patients with comprehensive care and support them through the different stages of their lives.

Along this journey of personal and professional discovery, his passion for reading and music continued to light up his life. Books and melodies provided constant refuge, inspiration and windows to other worlds, which only continued to nourish his spirit as he embarked on his career as a physician.

Reunion with a friend

While awaiting surgery for his cardiac arrhythmia, José suffered a sudden syncopal episode associated with his heart condition, temporarily losing consciousness and collapsing. One day, after getting up from a nap, he dressed in the dim light of his bedroom. What happened next is a bit of a blur, but he somehow lost consciousness and collapsed to the floor.

José is not sure how long he was unconscious, though it was probably less than a minute. Nevertheless, everything recounted in the following pages occurred within that span of time. It must have been brief, because if his cardiac arrest had lasted over a minute, he might have been left with rather obvious neurological sequelae, which was not the case. Based on his medical experience and what he went through, he is convinced that he suffered a cardiac arrest. However, since he was not monitored by an ECG and it happened at home, it was technically recorded in his medical notes as syncope (fainting) with loss of consciousness.

According to our concept of time, everything that José describes could not possibly have happened in under a minute. Here it is essential to understand that NDEs do not conform to our conventional model of space-time, as they unfold in another dimension. As we have seen in this book, quantum mechanics explains this with objective, evidence-based facts.

After he fell, José next remembers seeing himself from a strange perspective. He was standing and observing his own body, which was semi-reclined and inert in a corner, wedged between a piece of furniture and the wall. Despite the faint light filtering through the closed blind, the scene was sharp and clear. From his vantage point he felt perfectly fine, with no pain or discomfort.

Suddenly, everything was enveloped in a turquoise blue vortex. When he regained his senses, he was in a dark and boundless space, with a bright white light shining above. A deep and oscillating hum created a strange but not unsettling atmosphere.

He then hurtled towards the light. In what felt like no time at all, he found himself in an immense, luminous space, as if emerging from a dark cave to a mountaintop on a sunny day. His mind was completely clear and his senses were heightened, though they were hardly necessary in such a pure and serene environment.

A momentous feeling of peace flowed through him, imbuing him with inner and outer well-being. Around him stretched a rural, mountainous landscape, as beautiful as an alpine meadow in spring. The temperature was perfect, the light was radiant and the silence was broken only by celestial music that reminded him of ambient, New Age melodies or the compositions of Vangelis.

This landscape gradually turned whiter and more neutral, eventually dissolving into a radiant expanse that did not hurt his eyes. Sensing something behind him, he turned to his left and saw Tomás José, a dear friend who had died from a brain tumour in 1983.

Tomás José was wearing a dark, formal suit. He looked elegant and showed no sign of the scar and skull deformity left by his tumour operation. Most surprisingly of all, despite having been blind since the surgery, he now seemed to be able to see. However, he did not look José in the eyes, nor did he open his arms to hug him. His gaze communicated sadness and a clear message: 'Don't come. Continue on your way.' Though he did not utter these words, José knew he meant them just from the look on his face.

Knowing how much Tomás José had loved him, José understood that something important was happening. Despite his intense desire to embrace him, he held back, understanding that his loving friend was wordlessly conveying that he must carry on.

Aching with empathy

As he turned away from his friend, José noticed a dazzling white being to his right. Though towering and magnificent, it was in no way intimidating. It radiated an aura of infinite understanding, empathy and reassurance.

Suddenly, his entire life flashed through his mind. He witnessed not only the most significant moments, but also others that seemed trivial. It was not like watching a sequence of frames, but more like seeing a mosaic of images on a wall simultaneously – scenes he knew in the most exacting detail and could now appreciate parts of as a whole from his distant perspective.

What impacted him most was that he could feel the sensations and emotions of the people involved in each memory. He was not merely reliving what he had done; he was also acutely aware of the hurt he had caused others. This left him bewildered and stinging with remorse for the harm he had inflicted in the past out of ignorance.

The dazzling white being conveyed kindness and understanding to him, not with words, but as a direct transmission. It was not a line of reasoning, but a state of mind that told him: 'Don't worry. You didn't know. It's all right. Life is like that for everyone. Everyone does the best they can from their own state of being.'

He then noticed a hazy, undefined threshold before him. He was being lovingly invited to cross it, from a place of complete well-being. He felt so good that he would not have minded stepping over, but he was suddenly overwhelmed with guilt for all the pain he had caused. A cry broke out inside him and echoed throughout his being. He knew that he could not leave things as they were. He wanted to go back and make amends. It was as if he felt the need to cleanse his karma, and to do that, he needed more time. It was then that he understood what the look in Tomás José's eyes had meant: if he had embraced him, there would have been no going back, and his friend did not want that for him. Even today, his eyes well with tears when he recalls his act of immense generosity.

He looked at the being of light, which understood his desire to return. Wordlessly but lovingly, it told him: 'Do what you need to do. That is all right too.'

He started falling backwards and spun with vertigo, as if he were riding a rollercoaster. Then he felt that he was back on the floor, inside his body, just as he had seen himself at the start of his out-of-body experience. He felt weak and in pain. With great effort, he rose from the floor to lie back down on the bed. Clumsily and somewhat confused, he struggled to order his thoughts and to make sense of what had just happened.

Only gradually did he resume his ordinary life. It took him several months before he could share the experience with anyone. One summer night, during an after-dinner conversation and without any introduction. José told his story, without going into many details, for the very first time.

Who was Tomás José?

José met Tomás José in his local parish church in the late 1970s after he was asked to keep him company. When he was just 18, Tomás José had undergone surgery for a brain tumour, which left him with severe neurological sequelae, including blindness and significant weakness on the right side of his body. Despite this hardship, he maintained an unshakeable faith and a positive outlook towards life.

Over the years, the two forged a great friendship full of mutual affection. They even began to share their circles of friends, leading to a great period of happiness in both of their lives.

Tomás José's health improved remarkably during this time, a testament to the positive impact of José's presence in his life. It was as if their friendship allowed his inner strength and potential to fully emerge. Four years later, however, the tumour returned. After many months of steady decline, the young man passed away.

Despite his misfortune, Tomás José always demonstrated a profound faith, which enabled him to accept and bear all the suffering caused by his illness with dignity. Even in his final moments, he remained strong and forward-looking. On the last day they spoke at length, he said to his friend, 'José Antonio, I'm praying for you.' This moved José deeply. Even now, 40 years later, he remembers Tomás José with fondness and deep admiration.

The days before the NDE

José had his NDE during one of the happiest periods of his life. He had managed to transfer to the outpatient clinic in La Ràpita, Tarragona province, in the Ebro Delta, a place he considered paradise, where he enjoyed his work and the life he had longed for with his family. The warmth of his colleagues and the feeling of having fulfilled a dream brought him great bliss.

His only concern was his heart. An arrhythmia that had worsened in the previous year was restricting his physical activity, leading him to start planning for surgery. He had never fainted or experienced cardiac arrest before, so the incident that triggered the NDE caught him completely by surprise.

There were two key moments in José's NDE.

The first was his poignant reunion with Tomás José, who avoided hugging him, but who also shared a message whose meaning José would only fully grasp later on.

The second was his empathic review of significant and trivial moments from his life. Most of these memories focused on his mistakes or acts of selfishness and he was not proud of them. While revisiting these moments, José experienced the physical and emotional sensations of the people affected by his actions, such as a knot in the stomach, shortness of breath, disappointment, sadness and anger. He later wondered why he did not also recall his acts of kindness, a question for which he still has no answer. This review caused him physical discomfort, despite having no body, and emotional distress. When he was invited to cross a threshold that he sensed was a point of no return, his overwhelming guilt and the need to set things right weighed more heavily on him. He was then told that he could return to his former life if he wished.

A radical life change

A few days after meeting José, with whom I had only spoken on the phone or exchanged text messages, I could tell he was a kind person. It came through in his choice of words and how he spoke. That is why I wanted to ask him how it was possible that he, who radiates such goodness, had an NDE in which he specifically recalled moments when he inflicted emotional pain on others, and not those occasions when he helped people, which I am sure vastly outnumber them. We also talked about how the NDE changed his daily life.

The change did not come abruptly, like a sudden flash of light. It took him several months to digest the experience, which plunged him into a deep inner turmoil, in what he himself called his 'dark night of the soul'. Even after he got his old life back, his joy had evaporated, though outwardly he did not show it. Little by little, he came to realize that making amends and helping others made him feel much better.

One day, an elderly woman came to his practice with a medical and social predicament similar to one that had afflicted his mother. In the past, he had failed to help his mother appropriately, and that mistake burned on his list of 'selfish acts'. This time, he took a personal interest in this woman and got involved in solving her situation as he should have done for his mother. Once it was all settled, he felt that his mother was 'smiling at him from the other side'.

He also contacted someone he had let down in his youth to apologize. The person downplayed the importance of the issue and they parted on good terms as friends. Shortly afterwards, this person became entangled in a complicated family-related mental health crisis and turned to José for expert advice. José got personally involved once again and was instrumental in resolving the situation. Though they likely never knew it, these people had helped him far more than he had helped them.

Telling his story

Six months after his NDE, after dinner, José decided to share what had happened. He was with his wife, his daughter and some friends – a close married couple and their adult daughter. They were all people he trusted completely; they had eaten a good meal and were enjoying the relaxed atmosphere of a summer night. He felt he was in a 'safe space', so, as they joked and laughed, he suddenly said: 'I'm going to tell you something that will surprise you, but it's true.'

When he began, they thought it was some kind of ghost story with a punchline, like just another joke. However, as he went on, they grew quiet and wide-eyed. After he finished, his wife broke the silence. 'I remember noticing the bump on your head. How could you hide this from me all this time?'

'I couldn't tell anyone,' he replied.

Still, this was exactly when José began to feel better. It marked a turning point in his recovery.

His experience transformed him into a more empathetic and intuitive doctor. Some patients noticed and mentioned it to him, but he would brush it off, saying, 'I'm just doing my job.'

One day, after he had first opened up about his experience, a woman who was devastated by the loss of her teenage son came to his practice. They spoke about death, the possibility of an afterlife and his own personal certainty after his NDE. The woman's face lit up with the prospect of reuniting with her son. She had never felt so comforted before. That was when José understood that giving visibility to this phenomenon was part of his new purpose in life.

He had reservations at first, fearing that it might damage his professional credibility, but he soon noticed that it did no such thing. Naturally, without him taking any initiative, people began to ask to interview him and to listen to him speak about his experience. Some have occasionally questioned the spiritual basis of his NDE, but never his sincerity as a person.

Inexplicable phenomena

In the first few months after his NDE, José encountered some strange electrical interference, such as flickering computer screens and light bulbs. He also noticed that objects would fall during emotionally intense moments when he was thinking about Tomás José. Once he understood these occurrences as proof of life from his friend, they stopped. They were no longer necessary. That is how he interprets them.

The scientific literature documents many inexplicable paranormal phenomena during NDEs, some of them externally verified. José finds this fascinating, given that they have no simpler, materialist explanation.

As a teenager, the future doctor had read Raymond Moody's book *Life After Life*. A few years before his own NDE, he came across a bestseller that he loved, *Al otro lado del túnel* (On the Other Side of the Tunnel), by the well-known psychiatrist José Miguel Gaona. He wants to make it clear that he is a voracious reader of non-fiction and reads about all kinds of different topics out of intellectual curiosity.

However, after his NDE, José became obsessed with understanding what had happened to him. He sought out books and watched every testimony he could find on YouTube. This gave him knowledge, but not inner peace. What healed him was sharing his own experience and receiving expressions of gratitude for nurturing hope that our consciousness will endure beyond our physical death and that we will reunite with our loved ones. Although he had never particularly feared death, he now has no fear of it at all.

Having always enjoyed his work as a family doctor, he had previously thought he would keep working past retirement age if he was physically well. Medicine is very rewarding, and José credits it for much of the good that has happened in his life.

Yet everything changed after his NDE. José lost the motivation to keep up with the continuous education and dedication the profession requires and yearned to delve into spirituality, support people and spend time with those closest to him. Therefore, he retired early at the age of 63.

Today, José spends his time pursuing his insatiable curiosity about spirituality, wisdom literature, mystical and paranormal experiences and meditative practices like zen and mindfulness. Having come to terms with his NDE and aware of the profound responsibility that a second chance at life entails, he now provides private support and guidance free of charge. To this end, he has created the website www.doctormorales.es, where those interested can learn more about his NDE and watch some of the many interviews he has given on different YouTube channels.

CONCLUSION

Science cannot solve the ultimate mystery of nature.
And that is because, in the final analysis, we ourselves
are part of the very mystery we are trying to solve.
Inspired by a quote by Max Planck

The conclusion is a vital and highly practical chapter of any rigorous investigation. Studying the profound and surprisingly common phenomenon of NDEs, first via the scientific method and then from a quantum perspective, provides crucial insight into the fundamental structure of human beings and leads to a profound understanding about the nature of our existence.

A surgeon's training is based on the Cartesian and Newtonian scientific method, a rational approach grounded in observation and experimentation. Its goals are to cure or heal, to alleviate suffering when that is not possible and to always comfort the patient and their family.

When faced with any organic or psychological phenomenon, we seek to answer two fundamental questions: what is the cause, known as the aetiopathogenesis, and what is the mechanism of its development, or pathophysiology.

The conventional scientific method is rationalist, materialist, deterministic and inherently sceptical. It considers matter to be the basic building block of the universe and operates within the macroscopic world.

From this scientific standpoint, human beings are made entirely of matter. Both the body and the mind, with all its psychological functions, are an epiphenomenon of the brain. Physical death, as René Descartes famously implied with the words, 'I think, therefore I am', would therefore mean the end of our existence. According to the second law of thermodynamics, everything tends towards entropy and disorder. Death causes the total breakdown of the organism and the cessation of its functions. When thinking stops, existence ends. Our local, neuronal consciousness disappears.

Patients diagnosed with clinical death present with cardiac arrest, respiratory arrest, areflexia (the absence of reflexes) and loss of consciousness, while their EEG trace is flat, showing an isoelectric line within 15 seconds. As mentioned before, sometimes we can revive these patients by performing CPR. Some report having what we call NDEs while in this state. The assessment and study of these NDEs form the basis of this book.

Conventional scientific thought diagnoses NDEs as hallucinations caused by a lack of blood flow, to which the brain is extremely sensitive. As we have seen, a shortage or absence of blood and oxygen in the brain causes anoxia, hypercapnia, hyperkalaemia and metabolic acidosis. These conditions swiftly lead to cellular damage: neuronal synapses fail, proteins are altered and cell membranes are destroyed, resulting in a disastrous mineral transfer and culminating in neuronal death within minutes.

The difference between NDEs and hallucinations

A rigorous study of NDEs reveals clear and significant differences when compared with hallucinations.

Clinical assessments of NDEs reveal a coherent and patterned structure. Their core features are consistently reported by different patients, varying in nuance but not in their fundamental nature. NDEs have a profound psychological impact on patients, who remember them vividly for life. This impact changes their existential outlook and how they view their purpose in life. The clinical profile of hallucinations is entirely different, contradicting every key aspect of NDEs.

NDEs also possess characteristics absent in hallucinations that conventional science cannot explain. These include the transfer of information independent of space and time and the ability to pass through solid structures.

By performing an fMRI scan on a patient while they recount their NDE, we can observe the activating brain regions. One of the most striking findings is occipital lobe activation when the patient describes emotionally impactful images they saw during their NDE. This demonstrates memory activity linked to mirror neurons and provides strong evidence that the patient is not lying and truly recalls seeing these images from a perspective outside their inert body while CPR was being performed.

The differences between hallucinations and NDEs are clear enough to suggest that other, unknown phenomena are involved in NDEs that lie beyond the scope of the scientific method.

The quantum approach

The development of quantum physics from the 19th century onwards provided a new conceptual framework for consciousness. This discipline studies the microscopic world. The interpretation of its laws and principles ushered in a new paradigm, requiring a completely new way of thinking.

Applying quantum mechanics to a structural interpretation of human beings opens new possibilities for understanding transcendent phenomena, especially NDEs.

As matter, the body is low-frequency energy. The quantum interpretation of matter holds that it is structurally empty, with vast spaces between subatomic particles. This brings to mind patients' accounts of passing effortlessly through solid structures.

The quantum view of our mental activity is that it is also presented as energy, specifically in the form of high-frequency electromagnetic waves.

Ultimately, we must accept the existence of a non-local consciousness or Supraconsciousness: a high-frequency energy imperceptible to our senses. This is our true identity and makes us unique and unrepeatable. Capable of collapsing energy into matter, it is holistic with the primordial universal energy and its properties of omnipresence, omniscience and omnipotence.

This new existential vision of humankind raises profound questions about our most basic experiences and the nature of life and death.

Post-NDE changes

People who have had an NDE typically undergo a profound change in their approach to four chief concepts:

1) **The first is life.** Living beings can self-replicate and carry out biological functions involving the exchange of matter

and energy. Patients who have had an NDE view life as an amazing gift that must be lived to the fullest. This is a marvellous opportunity in our evolution. Being born means entering a body, for in essence we are stardust, and borrowing it for a limited time. Each experience in life is a lesson that fosters growth. If what we value in life springs from our materialistic identity, the non-self or ego, we will only perceive limitations and fears. The ego both chains us to the past, which torments us with guilt, and goads us into the future, shadowed with uncertainty and anxiety about achieving material goals. To value life through Supraconsciousness is to focus on the present moment, the only existential reality, where the dualistic separation between observer and object dissolves. Locality vanishes along with it, freeing us to enjoy the simple and beautiful essence of nature. As Newton said, 'Truth is ever to be found in simplicity, and not in the multiplicity and confusion of things.'

2) **The second is happiness.** Patients who have had an NDE are fully aware that authentic happiness is only achieved via Supraconsciousness. It is a state of peace, equilibrium, joy, stillness and silence, a pure experience of presence. Living with our materialistic identity, with the ego, we are caught between opposites. We cannot possibly eliminate the unwanted opposite, since one creates the other. Opposites are an illusion of the ego and there is always a struggle involving strain and anguish at the boundary between them. In reality, the ego does not provide happiness, but merely fleeting pleasure. The ego grounds

its identity in the material body. Knowing that the body is finite and will die, it is plagued by tension, anxiety and unhappiness.

Each psychological phenomenon has a neurochemical basis and operates through neurotransmitters. From this perspective, the ego's pursuit of pleasure is driven by dopamine, which stimulates reward pathways. In contrast, the authentic happiness of Supraconsciousness reflects a serene balance linked to the complex and regulatory functions of serotonin.

3) **The third is freedom.** There can be no true freedom without Supraconsciousness. As long as our mental dynamics are governed by the ego, the decisions we make depend on the architecture of our character and our ingrained habits, which shape our thoughts and feelings.
4) **The fourth is death.** Our society has a horror of death, treating it as taboo. We must completely transform this view: death is not the opposite of life, but a part of it. From the moment we are born, we slowly begin to die. Life is a preparation for death. Our dread of death alienates us from the truth of our existence, which is eternal. In death, we return to our origin.

   As long as we consider death the extinguishing of our existence, we will never live in peace and harmony, but in fear. For ultimately, at its root, all fear is a fear of death.

Death is not the opposite of life, but part of our existence.

Death and Supraconsciousness

By mastering the ego, we discover Supraconsciousness, our eternal existential reality. The primary and most profound duality, the opposition between life and death, then vanishes, and all other dualities fade away with it.

We cannot enjoy life, love life or even be genuinely happy if we see death as the enemy. Our awareness of life's brevity stems from the fear of death. Only by accepting death as one more waystation along our journey can we shake off our dread of time and fully inhabit the present moment.

When we love life, our expanded heart shrinks our fear of death. We must see our existence as a polarity with two extremes of life and death.

In death, we return to a state of rest. The circle continues to turn.

Death means the end of the ego and local consciousness.

We die as we have lived. When we discover Supraconsciousness, we lose our fear of death, secure in the knowledge that our true existence is eternal.

The ego spurs such a fierce attachment to the body that we cling to it even when it is ill and incapacitated. This is simply the fear of the unknown. Death lifts away the burdens of life, liberating us from our pains, illnesses and hardships, while granting us peace and harmony.

We can compare our fear of death to the fear of the dark.

Pain and the fear of death are often primarily due to the ego's self-centred psychological attachment to the body, rather than to any physiological suffering.

When we are born, Supraconsciousness enters our body, identifies with it and forgets its true nature, which is holistic with the primordial energy. Those who discover and become aware of their existential reality see the body as an imposed and confining, though necessary, vessel for their evolution.

To shed our fear of death, we must live in the world without belonging to it. We must act on the world's stage without being defined by the role we have been assigned to play.

Final thoughts

Meditation prepares us to detach from the body. It allows us to rise up from ignorance and comprehend our existential reality.

Studying NDEs also teaches us about our existential reality: we are an undying spirit, clothed in a body for a blip in time – stardust that returns to the universe upon our death.

Patients who have had an NDE have glimpsed the true reality and do not wish to return to our three-dimensional realm. The joy, peace, happiness and love they felt during their NDE explain their reluctance to return. These individuals report that life on the terrestrial plane is rife with obstacles and manoeuvring. The lesson is clear: suffering is a condition of the here and now.

Quantum mechanics scientifically demonstrates that we are energy, revealing a certain parallel with the transcendental experiences of NDEs.

When we are born, everyone smiles with joy as we cry. When we die, everyone weeps as we find peace and bliss upon merging with Supraconsciousness.

Death as we know it does not exist; it is not a biological process, but a spiritual one.

The main conclusions drawn from the study and assessment of NDEs, grounded in scientific principles, help us to understand death and what lies beyond:

- In our structural conception of human beings, we distinguish three constituent parts: the body, local (or neuronal) consciousness and non-local consciousness (or Supraconsciousness).
- Supraconsciousness is a subtle, two-dimensional energy that is non-dual and non-local.
- There is a primordial or universal energy that is the intelligent designer of the universe and the foundational principle of all religions.
- Supraconsciousness is holistic with this primordial consciousness and shares its properties of omnipresence, omniscience and omnipotence.
- In physical death we slough off our bodily vessel, but our existential reality, Supraconsciousness, endures forever.
- There is no need to fear death. It is merely a transition to a higher dimension.
- Quantum biology demonstrates that quantum mechanics operates in living beings, revealing a clear parallel with the phenomena described by patients during their NDEs.
- There is sufficient evidence that NDEs are real. Patients are not fabricating their descriptions.
- NDEs allow us to understand our true existential reality.

Life is a game played on a three-dimensional board where we may gradually purge our impurities before cycling back to the primordial consciousness. Once we realize that Supraconsciousness is the true foundation of our existence, our domineering ego will no longer be able to debase us into suffering, illness, injustice and war. People who have had an NDE – perhaps the universe's gift of a second chance to embrace this primordial consciousness – attain lives of deep fulfilment, marked by connection, harmony, peace, health and a reverent love for the planet, for nature, for animals and for our fellow human beings.

Scan the QR code below to visit Dr Manuel Sans Segarra's website.

Song composed for Dr Manuel Sans Segarra's lectures and videos by Jan, a musician and composer.

BIBLIOGRAPHY

Alexander, E. (2013). *Proof of Heaven*. London: Piatkus. (*La prueba del cielo*. Barcelona: Zenith.)

Barnaby, B. (2008). *Beyond the Secret*. Mumbai: Pentagon Press. (*Más allá del secreto*. Barcelona: Robinbook.)

Blanch Matute, MA (2020). *Registros akáshicos*. Madrid: Oberon.

Bohm, D. (1980). *Wholeness and the Implicate Order*. London: Routledge. (*La totalidad y el orden implicado*. Barcelona: Kairós.)

Bohr, N. (1958). *Atomic Physics and Human Knowledge*. New York: John Wiley & Sons. (*Física atómica y conocimiento humano*. Madrid: Aguilar.)

Bolloré, MY & Bonnassies, O. (2025). *God, the Science, the Evidence*. New York: Abrams. (*Dios. La ciencia. Las pruebas*. Madrid: Funambulista.)

Bush, NE (1991). 'Is ten years a life review?'. In *Journal of Near-Death Studies, 10,* pp. 5–9.

Byrne, R. (2006). *The Secret*. New York: Atria Books. (*El secreto*. Madrid: Urano.)

Capra, F. (1975). *The Tao of Physics*. Boulder: Shambhala Publications. (*El tao de la física*. Málaga: Sirio.)

Clegg, B. (2017). *The Quantum Physics Bible: The Definitive Guide to 200 Years of Subatomic Science*. Richmond Hill: Firefly Books. (*La biblia de la física cuántica. Guía de viaje a través de 200 años de ciencia subatómica*. Madrid: Gaia.)

Dalai Lama. (2005). *The Universe in a Single Atom: The Convergence of Science and Spirituality*. New York: Broadway Books.

Damásio, A. (1994). *Descartes' Error.* New York: Putnam. (*El error de Descartes.* Barcelona: Crítica.)

Dawkins, R. (1986). *The Blind Watchmaker.* Harlow: Longman. (*El relojero ciego.* Barcelona: Tusquets.)

Descartes, R. (1994). *Tratado del hombre.* Barcelona: Andreu (Grupo Roche).

Einstein, A. (1930). 'Religion and Science'. In *New York Times*, pp. 1–4.

Einstein, A. (1935). *The World as I See It.* London: The Bodley Head. (*Mi visión del mundo.* Barcelona: Tusquets.)

Fenwick P and Fenwick E. (2008). *The Art of Dying.* London: Continuum. (*El arte de morir.* Girona: Atlanta.)

Feynman, RP. (1992). *Surely You're Joking, Mr. Feynman?* New York: Vintage. (*¿Está usted de broma, Sr. Feynman?* Madrid: Alianza.)

Fuster, V & Corbella, J. (2016). *The Science of a Long Life.* New York: Hearts of Our Children. (*La ciencia de la larga vida.* Barcelona: Planeta.)

———. & Sampedro, JL. (2009). *La ciencia y la vida.* Barcelona: Plaza y Janés.

Gaona Cartolano, JM. (2015). *El límite.* Madrid: La Esfera de los Libros.

———. (2022). *On the Other Side of the Tunnel.* Independently published. (*Al otro lado del túnel.* Madrid: La Esfera de los Libros.)

Goswami, A. (2013). *The Physics of the Soul.* Newburyport: Hampton Roads. (*La física del alma.* Barcelona: Obelisco.)

———. (2006). *The Visionary Window.* Wheaton: Quest Books. (*La ventana del visionario.* Madrid: La Esfera de los Libros.)

———. (2008). *Creative Evolution.* Wheaton: Quest Books. (*Evolución creativa.* Madrid: La Esfera de los Libros.)

———. (2018). *The Everything Answer Book.* Newburyport: Hampton Roads. (*El libro que lo responde todo.* Barcelona: Obelisco.)

Greyson, B. (1983). 'The Near-Death Experience Scale: Construction, Reliability, and Validity'. In *Journal of Nervous and Mental Disease, 171*(6), pp. 369–375.

———. (2000). 'Dissociation in people who have near-death experiences: out of their bodies or out of their minds?'. In *The Lancet, 355*, pp. 460–463.

Grof, S. (2006). *The Ultimate Journey: Consciousness and the Mystery of Death*. San Jose: Multidisciplinary Association for Psychedelic Studies. (*El viaje definitivo: la consciencia y el misterio de la Muerte*. Barcelona: La Liebre de Marzo.)

Hameroff, SR & Penrose, R. (2014). 'Consciousness in the universe: a review of the "Orch OR" theory'. In *Physics of Life Reviews, 11*(1), pp. 39–78.

Hawking, S. (2001). *The Universe in a Nutshell*. London: Bantam. (*El universo en una cáscara de nuez*, Barcelona: Crítica.)

———. (2008). *The Theory of Everything*. Mumbai: Jaico Publishing House. (La teoría del todo. Barcelona: Debolsillo.)

Hawkins, DR. (2006). *The Eye of the I*. Carlsbad: Hay House. (*El ojo del yo*. Barcelona: Obelisco.)

Heisenberg, W. (1971). *Physics and Beyond: Encounters and Conversations*. London: HarperCollins. (*Más allá de la física. Atravesando fronteras*. Madrid: Biblioteca de Autores Cristianos.)

Jung, CG. (1994). *Simbología del espíritu*. México, D.F.: Fondo de Cultura Económica.

Kaku, M. (2014). *The Future of the Mind*. New York: Doubleday. (*El futuro de nuestra mente*. Barcelona: Debate.)

Kandel, ER. (2007). *In Search of Memory: The Emergence of a New Science of Mind*. New York: WW Norton & Co. (*En busca de la memoria. El nacimiento de una nueva ciencia de la mente.* Madrid: Katz Editores.)

Kübler-Ross, E. (2003). *On Death and Dying.* New York: Simon & Schuster. (*Sobre la muerte y los moribundos.* Barcelona: Debolsillo.)

———. & Kessler, D. (2014). *Life Lessons.* New York: Simon & Schuster. (*Lecciones de vida.* Barcelona: Grup Editorial 62.)

Lao Tzu. (2000). *Tao Te Ching.* London: Penguin.

Lázaro, C. (2018). *Lo que dicen los expertos sobre las experiencias cercanas a la muerte,* Almería: Guante Blanco.

Long, J & Perry, PJ. (2011). *Evidence of the Afterlife.* San Francisco: HarperOne. (*Evidencias del más allá.* Madrid: EDAF.)

Miret, S. (2019). *Biología cuántica.* Madrid: Siglo Actual Libros.

Moody, RA. (2001). *Life After Life.* London: Rider. (*Vida después de la vida.* Madrid: EDAF.)

———. (1995). *Más sobre la vida después de la vida.* Madrid: EDAF.

Morse, M. (1993). *Transformed by the Light: The Powerful Effect of Near-Death Experiences on People's Lives.* London: Piatkus Books. (*Transfigurados por la luz: el poderoso efecto de las experiencias cercanas a la muerte sobre nuestras vidas.* México D.F.: Diana.)

Newton, M. (2002). *Journey of Souls: Case Studies of Life Between Lives.* Saint Paul: Llewellyn.

Paolelli, E. (2015). *Neurocuántica. La nueva frontera de la neurociencia.* Barcelona: El Grano de Mostaza.

Parnia, S, *et al.* (2014). *The Lazarus Effect.* London: Rider.

———., *et al.* (2014). 'AWARE–AWAreness during REsuscitation: a prospective study'. In *Resuscitation, 85*(12), pp. 1799–1805.

Penrose, R. (1994). *Shadows of the Mind: A Search for the Missing Science of Consciousness.* London: Vintage. (*Las sombras de la mente. Hacia una comprensión científica de la consciencia.* Barcelona: Crítica.)

Plum, F & Posner, M. (1983). *The Diagnosis of Stupor and Coma.* Philadelphia: FA Davis.

Punset, E. (2006). *El alma está en el cerebro.* Madrid: Santillana.

Romero, T. (2020). *24 Minutes on the Other Side.* Independently published. (*24 minutos en el otro lado. Vivir sin miedo a la muerte.* Vergara.)

———. (2022). *Spirits in the Bonfire.* Independently published. (*Espíritus en la hoguera.* Ediciones de la Parra.)

———. (2023). *Los animales también van al cielo.* Ediciones de la Parra.

Rossner, M. (2013). *¿Tienes una invitación para ir al cielo?* Barcelona: Urano.

Sabom, M. (1998). *Light and Death: One Doctor's Fascinating Account of Near-Death Experiences.* Grand Rapids: Zondervan.

Sartori, P. (2015). *ECM. Experiencias cercanas a la muerte.* Barcelona: Kairós.

———. (2016). *What Is an NDE? A Beginner's Guide to NDEs.* London: Watkins Media.

Schrödinger, E. (2012). *What Is Life?* Cambridge: Cambridge University Press. (*¿Qué es la vida?* Barcelona: Tusquets.)

Selbie, J. (2021). *The Physics of God.* San Francisco: New Page Books. (*La física de Dios.* Málaga: Sirio.)

Swedenborg, E. (1993). *Awaken from Death.* San Francisco: James F. Lawrence.

Talbot, M. (1996). *The Holographic Universe.* London: HarperCollins. (*El universo holográfico.* Madrid: La Esfera de los Libros.)

Tolle, E. (2005). *A New Earth.* London: Viking. (*Un nuevo mundo, ahora.* Barcelona: Debolsillo.)

Van Lommel, P. (2011). *Consciousness Beyond Life.* San Francisco: HarperOne. (*Consciencia más allá de la vida.* Girona: Atalanta.)

———, *et al.* (2001). 'Near-death experience in survivors of cardiac arrest: a prospective study in the Netherlands'. In *The Lancet, 358*(9298), pp. 2039–2045.

Versyp, T. (2010). *La dimensión cuántica. De la física cuántica a la conciencia,* Huarte: Ulzama.

———. (2022). *Coherencia cuántica y vida.* Huarte: Ulzama.

Vigne, P. (1994). *¡Cierto! Hay vida más allá de la Muerte.* Barcelona: De Vecchi.

Weiss, BL. (1997). *Through Time into Healing.* London: Piatkus. (*A través del tiempo.* Barcelona: Ediciones B.)

———. (2018). *Many Lives, Many Masters: The True Story of a Prominent Psychiatrist, His Young Patient and the Past-life Therapy That Changed Both Their Lives.* London: Piatkus. (*Muchas vidas, muchos maestros. La historia real de un psiquiatra, su joven paciente y la terapia de regresión que cambió sus vidas para siempre.* Barcelona: Debolsillo.)

Wilber, K. (1993). *Vida después de la Muerte.* Barcelona: Kairós.

———. (2001). *No Boundary.* Boulder: Shambhala Publications. (*La conciencia sin fronteras.* Barcelona: Kairós.)

———. (2001). *Quantum Questions.* Boulder: Shambhala Publications. (*Cuestiones cuánticas.* Barcelona: Kairós.)

Xu, G; Mihaylova, T; Li, D; *et al.* (2023). 'Surge of neurophysiological coupling and connectivity of gamma oscillations in the dying human brain'. In *Proceedings of the National Academy of Sciences of the United States of America (PNAS), 120*(19), University of Michigan.

Yates, J. (2017). *The Mind Illuminated.* Carlsbad: Hay House. (*La mente Iluminada.* Málaga: Sirio.)

Yogananda, P. (2023). *Autobiography of a Yogi.* Los Angeles: Self-Realization Fellowship. (*Autobiografía de un yogui.* Barcelona: Paidós.)

INDEX